# HEARTH AND HOME

## WOMEN AND THE ART OF OPEN-HEARTH COOKING

# HEARTH AND HOME

## WOMEN AND THE ART OF OPEN-HEARTH COOKING

### FIONA LUCAS

JAMES LORIMER & COMPANY, LTD.
TORONTO

*I would like to thank* Lynn Schellenberg, Acquisitions Editor at James Lorimer & Company, for stalwartly sticking with me, and for so many suggestions to improve the text and match it with the photographs; Vince Pietropaolo for the wonderful action shots taken at Black Creek Pioneer Village; Derek Cooke, Curator of Black Creek Pioneer Village, because he gave so much time so willingly and participated cheerfully; and all the women who appear in the photographs.

Copyright © 2006 by Fiona Lucas

All rights reserved. No part of this book may be reproduced or transmitted in any form or by any means, electronic or mechanical, including photocopying, or by any information storage or retrieval system, without permission in writing from the publisher.

James Lorimer & Company Ltd. acknowledges the support of the Ontario Arts Council. We acknowledge the support of the Government of Canada through the Book Publishing Industry Development Program (BPIDP) for our publishing activities. We acknowledge the support of the Canada Council for the Arts for our publishing program. We acknowledge the assistance of the OMDC Book Fund, an initiative of Ontario Media Development Corporation.

**Library and Archives Canada Cataloguing in Publication**

Lucas, Fiona
    Hearth and home : women and the art of open-hearth cooking / by Fiona Lucas.

Includes bibliographical references and index.
ISBN10: 1-55028-921-7
ISBN13: 978-1-55028-921-3

    1. Fireplace cookery--History. 2. Hearths—Canada—History. 3. Women—Canada—Social conditions. I. Title.

TX840.F5L82 2006        641.5'8        C2006-900036-0

Illustration on Contents page: a brass skimmer from Black Creek Pioneer Village.

James Lorimer & Company Ltd., Publishers
317 Adelaide Street West, Suite #1002
Toronto, ON
M5V 1P9
www.lorimer.ca

Printed and bound in Canada.

# Contents

| | |
|---|---|
| Introduction | 7 |
| 1 Pioneer Hearths | 13 |
| 2 Prosperous Cooks and Kitchens | 23 |
| 3 City Cooks and Elegant Fare | 31 |
| 4 Cooking for a Living | 41 |
| Epilogue: From Wood Fire to Coal Fire | 49 |
| Bills of Fare | 51 |
| Sites of Interest | 56 |
| Notes | 62 |
| Selected Bibliography | 65 |
| Photo Credits | 69 |
| Index | 70 |

# Introduction

The kitchen was the main workplace of colonial Canadian women within family-centred households for generations. The great fireplace was each kitchen's focus, its heart. Its place at the centre of the family was expressed by pioneer settler Frances Stewart:

> *After a tedious walk [in the forest late at night] we at last got home, and found our dear old log house bright with a blazing fire and faithful Cartouche wagging his tail and jumping to welcome us home.*

At the fireplace, the daily cooking of meals was women's most regular year-round activity, but fire was essential for other culinary tasks too, such as rendering lard, preserving wild berries, and pickling home-grown vegetables. Most important, the fire boiled the water needed for the heavy household tasks of laundry and dishwashing—and for dyeing homespun woollens, boiling potatoes to extract their starch for laundry, for melting candle tallow, even for thawing snow for water when the well iced over. Hot water also warmed wooden churns in cold seasons to make the cream turn to butter more quickly, scalded tin pans to prevent soured milk, and bathed a newborn baby. And a fire meant the kettle could be boiled for a cup of tea to enjoy a quiet moment during a busy day or be poured into ceramic hot-water bottles for bedtime warmth. The kitchen fire's radiant heat served as furnace, clothes dryer, and mattress freshener. It slowly dehydrated herbs, legumes, and sliced apples and pumpkins for winter storage, revived sickly newborn lambs, and dried iron and tin to prevent rusting. Dishcloths hung from the mantel to dry by the fire. Smoke rising up the chimney helped preserve ham, pork, and eels. Firelight allowed everyone to do tasks before bedtime—knitting, hooking a rug, whittling a wooden bowl, mending a harness, endless darning—while someone read

*Above: A pan of fluted moulds for little cakes and pastries. Opposite: A hearth cook places a skewer of lamb into place for roasting in the tin kitchen.*

# Hearth and Home

*Many household and agricultural tasks could be accomplished at a firepit. Taking the cooking outdoors helped the family keep their cabin cool during the summer.*

aloud a poem, a bible passage, or a newspaper story to the assembled family.

In the earliest European settlements in the New World, men had to cook if they wanted to eat; but once their women arrived to share the rigours of the new life, few men cooked regularly in the family home. However, the men used fire for their chores on the farm, such as cauterizing an ox's wound and washing a sheep's wool. Often, they assumed responsibility for smoking the meat, fish, and game, as well as boiling the maple sap. In warm weather, women often abandoned the kitchen hearth for an outdoor firepit in which to cook, bake, and heat water. The firepit was also the place to undertake messier, smellier tasks such as drying flax and salt cod, dyeing homespun wool threads, soapmaking, and candlemaking. Autumn butchering required a fire, too.

Scratching out a living in the bush, pioneers wasted nothing. Even the fire's ashes were useful. For some farm families, a prime source of income was selling wood ash for the manufacture of soap, glass, and the leavening agent called pearlash. Many a pioneer housewife kept a barrel of ashes in the barn to use in soapmaking, as garden fertilizer, and to boil with corn to make hominy.

## OPEN-HEARTH COOKERY

Cooks bent or knelt to hearth level, reaching into the open fireplace, hence "down-hearth cookery" or "open-hearth cookery," though these terms only came into use with the invention of the raised hearth in the eighteenth century. Within the fireplace, pots hung from S-hooks or adjustable trammels, either on a pole fixed between both jambs (walls) or, after the mid-eighteenth century, from a crane that swung out from one jamb. Shovelfuls of glowing coals, carefully piled on the hearthstone, formed individual fires for three-legged, long-handled trivets, grills, skillets, and frypans. Three legs were sturdier than four on an uneven floor, and long-handled pots and implements distanced the cook

## Introduction

from the fire, so she did not roast along with the meat. To handle hot dishes, she used potholders of leather or thick woollen cloth, or tools such as pothooks.

The basic principles and technology of fireplace cooking changed little for generations —more innovation has happened in cooking in the past century and a half than in the preceding several centuries. Boiling, broiling, frying, roasting, and baking are the five techniques basic to all open-hearth cookery, despite regional variants.

Boiling was the easiest process, but it required finesse. Starting meat in cold water, for instance, ensured tenderness, but bagged puddings were plunged into rapidly bubbling water. Pork chops and beef steaks were broiled directly on a greased iron grill. Eggs, fish, fritters, sausages and onions, pancakes, and doughnuts were all pan-sizzled over glowing coals. Roasting techniques could be as simple as hanging a chicken by a twisted string from a nail in the mantel, or as complicated as skewering a great sirloin on a horizontal iron spit, which rotated via a wheel and pulley attached to a spring-driven clock jack on the wall. Spit-roasting was considered the monarch of cooking techniques.

A common method for the fifth technique, baking, was to place a lidded iron pot on a coal pile and then heap it over with more coals, plus a blanket of ashes. A temporary oven could be rigged by turning a hot iron pot upside down over a pan placed directly on the hearth. A griddle was ideal for quick-baking

*Above: By heaping hot coals under and over a bake kettle, colonial cooks could manage quite well without an oven. Below: Wafer irons were mostly square or round, so this heart shape was unusual.*

# Hearth and Home

*This well-equipped brick hearth at Kings Landing Historical Settlement includes a moveable crane, several iron pots, and both a tin kitchen and a brass clock jack for roasting meat. Adjusting heat levels was managed by shifting the pots back and forth along the crane or up and down on hooks.*

cookies, scones, pancakes, oatcakes, and crumpets. A wafer iron was a hinged utensil for baking fancy sweet crackers or waffles. Baking could also be done in a communal bake oven, or in the family's own bake oven built alongside the hearth in the fireplace, if the household had the space and the income to afford this relative luxury.

Fundamentally, iron hearth equipment was similar throughout the colony; it varied only in quality and quantity, according to the owner's financial means, ethnic and cultural traditions, and the ability of the womenfolk to adapt their cooking to different circumstances. Colonial cookware was predominantly iron, because it cost less than copper, brass, and bronze; most kitchens possessed several black iron pots. The Acadians called the basic stew or soup pot a *coquemar*, the Canadiens a *chaudière*, the English a *kettle*, the Americans a

# Introduction

*A wide range of cooking utensils hang at the ready behind this cook at Hutchison House Museum in Peterborough.*

*cauldron*. The ubiquitous lidded iron pot of the backwoods was a *bake kettle* to the British, but a *Dutch oven* to the Americans; a flat iron disk for quickly baking the wide variety of flat cakes was a *girdle* to the Scots, but a *griddle* to others. Iron was scoured clean with sand, pumice, or emery rock, even ashes and water, but not soap since it damaged the iron, then carefully wiped dry. All pots were heavy, so heaving them from crane to hearth to table required muscle.

Colonial housewifery necessitated not only physical muscle but an emotional stamina we can only strive to imagine. The pioneer housewife's contribution to the family's well-being was essential—"prosperity depends on female industry," wrote a New Brunswick settler, Frances Beavan, in 1845. Regardless of geographical and

## Hearth and Home

*A couple shop for produce at a farmer's market. Right: A wide variety of fruits and vegetables were planted by the early settlers.*

class differences, their ethnicity, their skills or knowledge, and strength or health, most women were the primary homemakers in early Canada, even if they were also farmers, fishers, innkeepers, teachers, weavers, kitchen employees, country shopkeepers, or businesswomen. Most women undertook the unrelenting preparation of family meals at their hearths, day in and day out, regardless of weather, season, illness, pregnancy, or family dynamics.

This narrative of Canada's early cooks and their hearths travels more or less chronologically and geographically across the developing colony, from Acadia on the Atlantic Coast in the seventeenth century, up the St. Lawrence River, as French, English, German, and American settlers moved inland into Quebec and then Upper Canada, followed by several British waves after the War of 1812. By the time Eastern Europeans were moving onto the Prairies in great numbers in the 1870s, the era of hearth cooking was waning, and indeed many settlers never used a hearth on the Prairies, so *Hearth and Home* is primarily a pre-Confederation and eastern story. As these earliest Canadian settlers travelled steadily westward seeking new land to colonize, they carried their cooking technology and food preferences with them, sometimes adopting Native ways, sometimes assimilating the dominant Anglo ways. But always, building a fireplace was an essential first project. The familiar image of the great kitchen hearth is a portal into the past, and to consider it is to delve into the shared history of Canada's women.

# 1 Pioneer Hearths

Top: A bake kettle.
Above: A mortar and pestle was indispensable to colonial cooks.

The cooks at the Habitation of Port Royal and at Ste. Marie among the Hurons, the earliest communities in New France, were, by default men, because the first women and families had not yet arrived. Canada's first European farmwife was Marie Rollet, who reached Hochelaga (Quebec City) in 1617 with her husband, apothecary Louis Hébert, their three children, and a servant. By 1663, Canada's population included one woman for every six men, but two decades later so many marriageable women—called *les filles du roi* (the king's daughters)—had been sent to the colony in a program of immigration that the numbers had equalized. Most were orphans accustomed to farm work, and within weeks they were wives on pioneer farms. In Acadia (now encompassing modern Nova Scotia, New Brunswick, and Prince Edward Island), the first women arrived in 1636 from the region south of the Loire River, while the first German emigrants established Lunenburg in Nova Scotia in 1753. After the American Revolution, five destitute Huguenot women

# Hearth and Home

*The pioneers' hearths devoured endless supplies of wood, which kept the men and boys busy with chopping and hauling chores.*

and their thirty-one children were the first refugees to reach Upper Canada, followed by thousands of Loyalist British, German, American, Pennsylvania-German, and African-American people, who spread into the backwoods of Nova Scotia, New Brunswick, and Upper Canada.

Having located a water source on their land claims, settlers pitched a tent or erected a shanty, dug an outdoor firepit, then commenced chopping down the trees for their first clearing. Soon, they built simple dwellings of one or two rooms, centred around the fireplace. The cooking was basic, with few utensils, limited ingredients, and a small repertoire of recipes.

## THE FIRST COLONIAL KITCHENS

Acadians lived in hamlets on wetlands converted into productive pastureland, and Canadiens, *les habitants*, cultivated small but mostly self-sustaining wheat farms along the St. Lawrence. Recalling their homelands of Normandy, Brittany, and Poitou, the houses of early French Canadians centred socially and architecturally around a hearth. Although differences existed between typical Acadian and Canadien rural homes, most were one- or two-room log or frame cabins, or, later, stone, with gable roofs, glass windows, and whitewashed interiors. As the family grew, so did the house, until three or four generations could live together in several linked rooms around the sparsely furnished main room.

This main room, *la salle*, was a large, all-purpose living space, focused on a primitive packed-clay-and-straw hearth, which was later replaced by a great fieldstone fireplace fitted with a swinging iron crane. A pair of iron fire-dogs supported the burning logs. Hanging from mantel pegs were perhaps a wrought-iron gridiron for broiling, a long-handled frypan, and various utensils, like a copper water-dipper, a brass skimmer, a wooden spoon, and an iron poker for adjusting the fire. On the mantel was often a crucifix, maybe a couple of candlesticks or a tin lamp of seal oil, and the essential mortar and pestle of stone, hardwood, or brass. A few shelves for stoneware jugs and vessels, a table with benches, a couple of rough pine chairs, maybe a rocking chair, a chest or armoire, and a cradle, were the likeliest pieces of furniture, and were often painted a reddish brown partially derived from beef blood. In winter, a weaving loom was located in one corner, with a spinning wheel nearby. Sometimes a storage loft was accessed by a ladder to one side of the fireplace, while a cool cellar stored bins of vegetables, baskets of eggs in sawdust, and barrels of salted cod. Hanging from the ceiling beams to keep them safe from mice were dried herbs, bags of beans, smoked meats, and sausage links. On the floor near the hearth might be a woven rug for the dogs to sleep on. The main table doubled as a kitchen work surface and eating place—once the white cloth

*This woman leans over her bubbling pot in a manner characteristic of hearth cooking.*

was laid, the pewter dishes set, and the chairs or benches drawn up. The man of the household sat at one end, and was always served first, while the children and farmhands sat on benches, unless the household lacked enough seats; if that was the case, his wife and children ate standing up.

The early cuisines of the two New France colonies were shaped by the Catholic calendar of numerous fasting and feasting days, familiar seasonal rhythms of scarcity and abundance, and the regional cooking of western France. The first colonists were astonished by the abundance of the forests and waters, to which they adapted their *bouillon* and *fricot*, hearty one-pot stews of legumes and such meats or fish as salt pork, rabbit, cod, and

# Hearth and Home

*Six corn cobs roast on a clever pronged utensil, while the cook stirs a stew in a pot hanging from the crane.*

shellfish, using heavy cast-iron *coquemars* and *chaudières* specially suited to slow simmering and tenderizing. Often, these dishes were topped with dumplings (*poutines* in Acadia). In France *bouillon* meant the flavoured broth created when cooking meats or vegetables or both, but it meant stew in New France. *Tourtières* were covered copper pans for *tourtes*, pies of either sweet fruit or savoury pork and pigeon, and a *terrine* was a rectangular earthenware baking dish. Roasting became commonplace, whereas previously in France only the wealthy could afford the wood and the specialized equipment the method required.

The earliest French settlers planted French seeds of oats, turnips, pease, lentils, onions, leeks, sorrel, lettuce, parsley, chives, and cabbage; transplanted their French apple trees and hop vines; and husbanded their French livestock into resilient Canadian survivors. They also planted Indian corn, beans, and squash together in rows of small hills in the Iroquoian way, as did later immigrants. From the Mi'kmaq of the Atlantic Coast and the Iroquois of the St. Lawrence, the French and British learned to tap maple trees, spear eels at night, roast pumpkins in ashes, and forage for such unfamiliar edible berries as blueberries,

## Pioneer Hearths

*Left: Boiling maple syrup over an open fire in the annual spring ritual. Right: By the light of a kerosene lamp, a cook slices a home-grown onion.*

cranberries, and chokecherries. The plentiful wild raspberries, strawberries, and cherries were similar to familiar European species. According to their particular traditions, all the women preserved both wild and orchard fruit as jam, with maple sugar, maple syrup or imported molasses, simmering the fruit mixture briskly enough to evaporate its moisture, but not enough to burn. Pickles were stored cold in stoneware jars and crocks covered with bladders, pieces of leather, or a thick layer of hardened melted fat. An old method of sealing a jam or marmalade was to cover it closely with a brandy-soaked paper circle.

Plentiful foodstuffs and fuel meant early Canadians usually ate better than their ancestors in their homelands. Colonial farmwomen evolved uncomplicated, flavourful cuisines unique to their new home, combining medieval memory, advice from the First Peoples, regional flora and fauna, and local growing conditions. In the mid- to late-eighteenth century, French-Canadian cuisine expanded to absorb English and American preferences, adding puddings boiled in cloths, tea and fortified wine, roast meat, and molasses and baking powders in cake baking. German and Mennonite cooking remained more distinct in its taste for sour and tart flavours.

### Backwoods Cooking

In her journal for autumn 1815, Nova Scotian Louisa Collins reported on the daily and seasonal housekeeping rhythms of her middle-class farming family. She, her six sis-

## Hearth and Home

*Settlers spent many summer hours picking berries to make preserves for winter. Left: A stoneware bottle used for preserves.*

ters, and their mother shared the haying and apportioned the female work. Louisa did most of the spinning and buttermaking. "We find the fireside the only cheerful companion when keen blows the north blast and heavy drives the snow," she wrote.

Lacking much in the way of equipment and ingredients, the meals eaten by the settlers during the first seasons, whatever their background or wherever they settled, were often primitive and limited, though even then results could be delicious because of the indefinable flavour that cooking over fire brings to simple ingredients. Vegetables could be buried in hot ashes to bake slowly, sometimes wrapped in protective leaves; potatoes were a classic example, but carrots, parsnips, beets, pumpkins, and squashes were done this way too, as was whole fish, "fit for the table of the most fastidious epicure," said Susanna Moodie. In the kettle, salt pork, dried pease, and wild herbs were blended into a basic stew, or Indian meal or oats were boiled for porridge. Cold sliced porridge was fried in bacon grease. Indian meal (cornmeal) blended with milk or water was the basis of many a plain cake, while bread and pudding was served with molasses as a sauce. A small pot would be kept warm on a flat stone beside the flames, and, if a coffee pot were lacking, coffee could be boiled in the frypan. Staples for most early settlers included pumpkins, maple molasses, flour and cornmeal for biscuits, wild game, dandelions for spring greens and for roasting as a coffee substitute, apples and berries. The Pennsylvania-Germans brought hogs and chickens with them, as well as cows, so they had a ready source of milk to make butter and the soft cheeses (*schmierkäse*) they enjoyed.

Two genteel Irish families, the Stewarts and the Reids, were the first white colonists to move into the unbroken bush around Peterborough, Upper Canada. Sailing the St. Lawrence in 1822, Frances Stewart remarked

## Pioneer Hearths

several times on the "charming" kitchen fires of long-settled Loyalist inhabitants, but their cheerful hospitality could not prepare her for her own backwoods housewifery. The Stewarts' one-room cabin was forty feet by twenty-eight feet—about twice the size of most—and the "large, light kitchen" had a "huge fire-place eight feet long." It was incomplete when Frances arrived, as she comments that "on waking in the [first] morning I saw the stars looking down through the aperture left for the chimney." She described the laying of the fire:

> We first put on a back log which is about a foot or eighteen inches in diameter and long enough to fill up the back part of the fireplace. Then we put in the dogs which you have seen I am sure in old houses, and on the dogs we lay smaller split sticks about five or six inches thick and the same length as the back log, then pile on chips and pieces of pine till we have it as high as we require, and you cannot think what a lovely pile it is, nor how cheerful it makes our little rooms in the evenings and mornings. Every evening before tea and every morning after breakfast we have a fresh back-log put on by one of the men and then we need only add smaller sticks to keep up a good fire. And in the morning we have only to take the kindled pieces out of the ashes, scrape the charred wood off the back log, put on fresh sticks and some chips, and in a few minutes we have a delightful fire which gives quite light enough all over

*Top: Wild rice was one of the foodstuffs indigenous to Canada that some settlers really enjoyed. Above: Sitting beside her fireplace in between preparing meals, this farm woman takes some time to spin.*

## Hearth and Home

*the room for dressing, sweeping, sitting and laying the breakfast table.*

There was a knack to getting the fire going, and everyone was concerned about a fire going out. Some fortunate families had a tinderbox. Before "lucifers" (sulphur matches) became available in Canada in the 1850s, it was the nightly ritual to insulate the hot, glowing coals under a layer of ashes before the family went off to bed. When the ashes were pulled aside the next morning, the coals were still hot. Not everyone did this successfully, and many a boy had to trot through the woods to request a coal from neighbours.

Wood for fuel was plentiful since in the earliest settlement years an important activity was razing the intimidating forests for the clearings that became the agricultural fields. For the most part, firewood was hauled on late-autumn or winter days, when farmwork was minimal, and then it was dried for several seasons. Cooking and heating with green wood was avoided by thrifty and sensible families. "The best wood for fires is hickory, hard maple, white ash, black birch, yellow birch, beech, yellow oak, and locust," advised *The Maple Leaf*, a Montreal magazine, in 1853. As far as the menfolk were concerned, a fireplace devoured endless piles of laboriously chopped wood. Frances Stewart's daughter Ellen later remembered the family's great kitchen hearth fondly, saying that the men "often brought home pine knots as the turpentine in them made a bright blaze and extra light for their work."

Frances Stewart did not write of the first breakfast cooked in the family's mighty fireplace, but a typical breakfast cooked over a fire in the backwoods, for both gentlefolk and the less privileged, may have been buckwheat pancakes with maple molasses and bacon with eggs fried in home-rendered lard, all washed down with water and peppermint tea or dandelion coffee. Mennonite families often included their soft milk cheeses and apple butter. Pulling up a stool, the mother or one of the eldest girls or the home help perched beside the fireplace to adjust the crane back and forth, flip the quick-baking pancakes on the hanging griddle with her knife, and tend the eggs and bacon sizzling down hearth in the skillet, while water for tea or coffee boiled in the copper kettle on the crane beside the griddle.

Just as a meal's many dishes could be cooking at once in the different areas of the hearth, the resourceful pioneer cook could use just one or two pots to cook many different dishes. Bachelor John Langton, an adaptable gentleman settler, managed to concoct a credible backwoods meal with one round-bellied pot, one frypan, one bake pan, and some utensils by boiling in succession potatoes,

*Above: A five-gallon butter churn of stoneware with cobalt blue decoration.*

## Pioneer Hearths

*Derby cakes, made with dried currants, bake on the hanging griddle, while lamb roasts in the tin kitchen, and sausages with onions sizzle in the spider, a long-handled frypan.*

beef, and a porcupine rice stew in the pot, then frying some pork and in its grease the potatoes in the frypan, and cooking a venison pie and then a cranberry tart in the bake pan. Some settlers were not as resilient. Letitia Hargrave, wife of the chief trader at the Hudson's Bay Company's York Factory depot, wrote: "I really think that the cooking here will end me. It is fearful." Many settlers wrote about bake kettles, the essential all-purpose three-legged iron pot of the backwoods that functioned as soup pot or pudding boiler when on the crane or as a miniature oven if sitting on the hearth. Preheating it on a pile of gentle coals before placing the bread dough or cake batter inside was important, as Susanna Moodie discovered: her "first Canadian loaf" was placed "unrisen . . . into a cold kettle," which was "heaped [with] a large quantity of hot ashes above and below." Alas, it was burned outside and raw inside.

Domestic skills, including the intricacies of cooking over fire, were already familiar to many of the British, American, African-American, European, and Scandinavian women who came to Canada, although they had to adjust to their new habitats. Little written material was available, so they learned by mutual assistance and lots of practice. As pioneers in an alien environment, they

# Hearth and Home

*A cauldron and kettle sit on the hearth's apron in front of two firedogs and a trivet.*

substituted indigenous ingredients; for instance, Indian corn (maize) replaced European corn (wheat). Gentlewomen, however, had much to learn in their new home.

American-born neighbours taught Mrs. Stewart and Mrs. Reid the basics of making bush yeast and bread, as well as hearth cooking, skills they as gentlewomen had not needed to cultivate, although like many of their class they knew how to bake, sew, and mend. In her turn, Frances Stewart taught others, like Catharine Parr Traill, who in 1854 wrote *The Female Emigrant's Guide*, an accumulation of female settlement wisdom. Gentlewomen all found themselves learning to cook competently through aggravating trial and error. This amusing anecdote is from diarist Mary O'Brien:

*I had just finished the first stage of my cooking and was about to shift my character from cook to gentlewoman when accidents began to happen. My little quarter of pork was dangling before the fire at the end of a skein of worsted, for having a loaf to bake I was unable to bake it as usual in the all-accomplishing bake kettle. I cast my eyes on the said bake kettle, and behold, its lid was raised upwards of an inch by the exuberant fermentation of the loaf within, which was threatening to run down its side into the ashes. Hastily then I was obliged to resume my labours and, seizing a knife, I cut from the top of the loaf the exceeding portion and placed it, much to my satisfaction, before the fire on a plate. There I hoped it would soon be converted into capital rusks. Of course the frying pan would have been the natural receptacle, but that was engaged in enacting dripping pan for the pork. Oh, who can number the uses and perfections of a Canadian bake kettle and frying pan.*

*I had just turned from the complacent contemplation of my arrangements when a treacherous stick, on which was resting at once for support and heat a saucepan containing a stew of cabbage and an old cock, gave way and my stew was emptied on my rusks. The rusks were spoiled, that could not be helped, but the lucky plate saved my old cock from being buried in the ashes and enabled me to restore my stew. Just then my guests arrived.*

# 2 Prosperous Cooks and Kitchens

After the initial settling in and accommodation to a new and challenging environment, diligent effort from all family members in clearing the woods, planting a farm, and building a barn brought greater prosperity to settlers within a few years. As soon as possible, they enlarged their

*The magnificent stone fireplace in the "first" Stong cabin at Black Creek Pioneer Village. The kitchen fireplace in the Stong family's "second house" contained a built-in bake oven, a sign of their prosperity. Top: A coffee mill.*

## Hearth and Home

*Top: The cook gives a quick stir to her pot, having pulled the crane closer to her from its place above the fire inside the fireplace. Bottom: Shifting the burning logs helps even out the heat and eliminate hot spots before the bread is put inside the oven.*

living quarters, either by adding rooms or building a new and substantial home. The commodious new house varied in regional architectural styles and floorplan, but it usually divided the living spaces into separate kitchens and parlours. The kitchen featured a large brick—or sometimes stone—fireplace that, for all but the moneyed class, denoted their domesticity. It was a workplace by day, a gathering place by evening. Since the parlour was generally reserved for special occasions, the kitchen fireplace remained the familial gathering place, unless it was truly the servants' domain only.

This second, or sometimes third, house represented a family's success in becoming colonial farmers or artisans, and also characterized the colony's growing strength in population, economic stability, and community expansion. Often, the houses developed into handsome homesteads, which included a summer kitchen, a smokehouse, a washhouse, and a bakehouse, all with separate fires.

Pennsylvania-Germans Daniel and Elizabeth Stong started out in 1816 in a twelve-foot-by-twenty-foot three-room log cabin with a magnificent stone fireplace that dominated the small main room. Elizabeth made her family's daily bread down hearth in the bake kettle. Then, a few metres away, in 1832, the Stongs built a spacious two-storey clapboard house with a smaller fireplace and a side oven of manufactured brick in the kitchen. Both houses, as well as the homestead's smokehouse, barn, and apple cellar,

## Prosperous Cooks and Kitchens

survive as the nucleus of Black Creek Pioneer Village in Downsview in north Toronto.

Many second, or expanded, houses had a bake oven of clay, brick, stone, or a combination, either as an adjunct to the indoor fireplace or outdoors under its own shelter. This usually showed a family had achieved prosperity, because only skill and expense built a good one.

To prepare the bake oven for several hours of activity, the cook laid a fire within it early in the morning. As it burned intensely, its flames licked the ceiling bricks, which absorbed the heat. After the fire had burned down and the glowing remnants of the logs were scraped out with an ash rake, the heat absorbed into the stone or brick would radiate for several hours onto the cakes, breads, and beanpots placed inside the oven's cavity. To judge an oven's

*Top: While the oven heats up, one cook sets the bread to rise and the other stamps out biscuit dough. Bottom: Using leather protectors, the cook tips the hot baked bread onto the table.*

# Hearth and Home

*First the bread goes deep into the oven, but quick-baking goods like biscuits and buns on tin sheets are placed in front for quick removal.*

heat before inserting the raw food, the cook threw in white flour or white paper to see how quickly it browned, or leaned in her full arm to see how long she could count before the heat became too uncomfortable on her skin. Each oven had its own idiosyncratic count, and each cook her own reading, but typically a ten- or twelve-second oven was judged ready to receive its first load of raw dough, adroitly placed into the heated cavity on a wooden paddle or peel. Bread was often baked directly on the oven's hot, ashy floor, while sweet batters, pastries, and fillings were poured into iron, ceramic, tin, or copper dishes. A door of iron or tin-lined wood plugged the opening, trapping the hot air, and was not removed until the smell of the baking was "right." Since a well-constructed oven retained its heat efficiently, bread was baked in the most intense heat, to be followed perhaps by several pies in shallow stoneware dishes; then, as the heat faded further, the oven could take a third load of quick-baking items like cookies, biscuits and buns, or dishes requiring slow heat, such as custards. The last vestiges of the warmth could be used to dry feathers for pillows.

Through the nineteenth century, inexpensive commercial goods for the household became gradually more available; for example, Americans began popularizing tinware for kitchen utensils, because it was durable, lightweight, and increasingly affordable. Tin rusted

## Prosperous Cooks and Kitchens

if not carefully dried; many a cook must have realized to her horror that a piece of her tinware had not been dried thoroughly. A well-equipped household had gill, pint, quart, and gallon jugs, and such useful items as funnels, colanders, candle boxes, and mouse traps, all made of tin. A reflecting oven or "tin kitchen" was an appliance used for roasting. A horizontal half-cylinder of tin encasing a spit, it had a handle for turning the spit projecting from one end. The tin kitchen was positioned on the hearth so that its open side faced the fire; a small door on the rounded back of the oven could be opened if the cook wanted to view the progress of the roasting.

Also made of tin was the common dishpan, which was placed on the table after meals and filled with hot water carried from the fire; when not in use it hung on the wall. Sometimes the dishpan was set into a dry sink under a window to take advantage of the natural light. When family resources allowed, a homemaker could request that a water pump be installed in her kitchen, but its cold water still had to be heated. Not everyone had this much; in 1852, one woman started housekeeping in Moose Factory, Manitoba, with "a large kettle, a frying pan and a teakettle, all of iron." By the 1850s, emigrant manuals often advised newcomers against bringing much kitchen equipment in their luggage, because kitchenware was readily available, either from general stores or artisans' shops. Once she was settled, a housewife could buy according to her husband's purse.

*Top: A tin mouse trap. Above: To take advantage of natural light, cooks often placed their dishpans on a window table.*

27

## Hearth and Home

*Top: Cranberries, eggs, cream, sugar and salt are ready to be combined into a delicious boiled pudding with a wine and sugar sauce. Bottom: After whisking the eggs and sugar together, the cranberries and cream are blended in, followed by the flour, until a soft and slightly sticky dough is attained.*

Even when a household became prosperous enough to hire some kitchen help, perhaps a neighbouring woman or one of the many women who came to Canada seeking a job, well-raised eighteenth- and early-nineteenth-century daughters, even those of gentlewomen, were expected to know at least one or two domestic skills from amongst baking, confectionery, mending, sewing, even ironing and spinning. Certainly many maintained a manuscript recipe book since most women, even those financially comfortable, spent time in the kitchen, if only to direct the cook. British North American cooks had a wide cookbook selection available to them, two being Hannah Glasse's *The Art of Cookery Made Plain and Easy* (first printed in 1747), the most successful eighteenth-century cookbook and still in use in the nineteenth century, and Eliza Maria Rundell's *New System of Domestic Cookery* (1806 through 1860), the most successful of the nineteenth century. Canada had two regional cookbooks: *La cuisinère canadienne* (1840) and *The Female Emigrant's Guide* (1854); both spoke to the realities of the Canadian housewife.

Other time savers and conveniences for the kitchen were introduced throughout the 1800s. By the mid-century, the average cook no longer needed to laboriously whisk multiple egg whites to a stiff froth to raise a single cake, because early baking powders to enhance leavening were readily available. Putting up jams and pickles became simpler and safer after the invention of the new glass

*Prosperous Cooks and Kitchens*

*Above: The ball of soft batter is carefully placed into a dampened and generously floured linen cloth. Right: Tied firmly with string, the pudding is ready to be carefully submerged in boiling water for a couple of hours.*

mason jars in 1858, although it was a long time before mason jars became common. One Prairie woman recalled that even early in the 1900s, "jams and jellies were stored in crocks, old pitchers, cups without handles, whiskey bottles with the neck cut off." West Indian white sugar, which came in the form of hard cones of various sizes, made superior fruit preserves, so once they could afford it, cooks preferred to use it instead of the local maple sugar, although pounding pieces of the cone to a powder added to their work.

The new kitchen aids and condiments were helpful, but cooking over fire still required a finesse that came only through experience. Successfully boiling a pudding, for instance, required attentiveness. Cookbook authors cautioned the cook to "take great Care that the Bag or Cloth be very clean, and not soapy, and dipped in hot Water, and then well flowered. If a Bread-pudding, tye it loose; if a Batter-pudding, tye it close; and be sure the water boils when you put the Pudding in, and you should move your Puddings in the Pot now and then, for fear they stick." Savoury and sweet puddings were an essential part of every meal for everyone, especially the British. They

## Hearth and Home

*Above: Many women maintained a manuscript recipe book. Right: White sugar came in the form of hard cones, which had to be broken, pounded and sifted into a powder.*

were either boiled, baked or steamed, and came in a huge variety of flavours and textures, such as Yorkshire pudding, beefsteak pudding, suet and herb dumplings, delicate egg-based custards, and sturdy Indian-meal puddings. In Canada, cranberries, blueberries and wild plums were new fruit varieties popularly added to puddings.

Country women shared the preparations for community and family events, such as church socials, Christmas gatherings, weddings, and harvest festivals. "Oxen and sheep were killed, and great roasts of beef and mutton hung on the spits before the open fireplaces. Roast beef, roast mutton, boiled potatoes and plum pudding were the staple fare at the wedding feasts," observed Elizabeth Norquay, a settler in the Red River area of Manitoba. For events such as barn-raising bees, grumbles might have been heard about the hard work needed to feed hordes of people, and the amount of alcohol consumed at such gatherings, but the women nonetheless took pride in providing their neighbours with tasty, filling and memorable meals. For days ahead of time, they baked dozens of pies and hams, stockpiled many eggs, and picked many berries. Then, on the appointed morning, the women rose especially early to rekindle the fire for the busy and long hours ahead of biscuit-baking and dishwashing ahead.

By the 1850s, as westward expansion was beginning, the pioneer communities of Eastern Canada had developed into prosperous communities. In 1867, the colonies of Nova Scotia, New Brunswick, Canada East, and Canada West confederated into the Dominion of Canada. Immigrants were transforming the agrarian Dominion into a thriving, multi-ethnic, industrializing young nation.

# 3 City Cooks and Elegant Fare

By the 1670s, a fully formed French colonial society of farmers, artisans, bourgeoisie, and elite had emerged in the New World. While the rural Acadians and Canadiens were adapting their medieval peasant cooking to pioneer conditions, the fashionably dressed bourgeoisie and government officials at Louisbourg, Montreal, and Quebec were drinking chocolate at breakfast and participating in up-to-date

*Top: A rolling pin for impressing patterns on hard cookie dough. Above: A servant with a coffee service in the grand drawing room of Old Government House in Fredericton.*

# Hearth and Home

*The potager stove started to appear in wealthier homes in the mid 1700s. In one kitchen at the Fortress of Louisbourg, an example stands beneath a set of hanging graduated copper saucepans. On the mantel is a mechanical jack attached to a spit for rotating meat in front of the roasting fire.*

Parisian dining practices, including the adoption of the new-fangled fork. Wives of seigneurs, merchants, government administrators, and military officers may have employed a cook and a housemaid or two, or had on staff enslaved Blacks or Natives, but some of them still spent part of their day in the kitchen, not just overseeing meal preparation but actually exerting themselves, too. Often, the lady of the house was the one who mixed the fancy almond macaroons and cheesecakes and twirled the molinillo (swizzle stick) in the chocolatière. Small though it was, French-Canadian high society boasted that it "dined as good as those in France."

Although the most common cooking facility was the fireplace, manor houses such as the governor's quarters and engineer's house at

## City Cooks and Elegant Fare

Louisbourg and the palatial residences of the governor, the intendant, and the bishop in Quebec City, had a *fourneau potager* (potager stove) with an in-built *paillase* (warming oven). A potager became indispensable to anyone trained in urban French cookery. It was a masonry box with a brick or iron hearth raised to waist height that featured several holes for saucepans; there was a door on the outside of the box through which the cook could lay a fire or load coals from the main fireplace. Sometimes the cook carefully shovelled small piles of coals onto the counter, and placed a footed *brazier* over them for grilling. From the cook's point of view, a potager was a marvellous invention, for not only was stooping and kneeling not required, but the contained heat permitted a precise control when one was cooking tricky cream sauces, sugar syrups, and ragoûts. In the mid 1700s, potagers—or hot hearths or stew holes, as the English called them—started appearing in private houses in New France, although, other than at a few military establishments and a few grand houses, they did not catch on in English Canada, nor with any other cultural group. Still necessary were a separate roasting and boiling fire and a bake oven. Since several fires burned simultaneously for much of the day, these well-equipped kitchens were hot, smoky, crowded, and smelly.

Generally, cooks in fine houses were women, but surprisingly, a fair number of male French chefs came to colonial Canada, either as émigrés or in the staffs of visiting dig-

*Top: A "fountain" of cold water, placed conveniently on a table, long before running water became standard in kitchens. Above: A silver chocolate service.*

## Hearth and Home

*Cakes, puddings, creams and jellies were often made in pretty fluted moulds for a fine presentation at table.*

nitaries. John Graves Simcoe, first lieutenant-governor of Upper Canada, and Mrs. Simcoe brought a French chef with them in 1793; his kitchen was a tent.

The early colonists' rough existence contrasted with the comfort experienced by the wealthier members of the seigniorial class and the vice-regal courts at Halifax in Nova Scotia and Newark in Upper Canada, where the bureaucrats considered themselves not colonists but temporary residents. Though surrounded by forest, they still aspired to be, and frequently managed to be, elegant. Governor and Lady Wentworth in Nova Scotia threw splendid parties that required intricate planning and great expense. "During the dancing there were refreshments of ice, orgeat, capillaire, and a variety of other things," such as macaroons, little iced cakes, spice cookies, individual cheesecakes, and tiers of glasses full of syllabub creams. The Simcoes both wrote extensive records of their five years in Upper Canada, from 1792 to 1797, in which "society" enjoyed a regular season of parties, concerts, and dances that included refreshments, buffet suppers, and formal dinners.

## City Cooks and Elegant Fare

One annual event was "Queen Charlotte's Birth Day Ball" in January.

The higher up the social ladder, the more refined the recipes and dining, the more use of such imported ingredients as almonds, spices, raisins, Parmesan cheese, lemons, wine vinegar, and olive oil, and the essential salt. French and especially Spanish wine was imported in large quantities too, and everyone drank it, often watered down, including the children. New France developed an internal trade in wheat, livestock, and such fruits and vegetables as onions, pears, and apples, while molasses, sugar, coffee, and rum arrived from the French West Indies, and salt cod went to Europe. Louisbourg received shipments of eggs from New England, and oysters went from Baie Verte in Acadia to Montreal.

A hired cook in a large town or country house worked in a much more lavishly equipped kitchen than her own at home; whereas her kitchen at home was the family's common workroom, the domestic offices of grand houses consisted of multiple rooms branching off the central kitchen. Sir Allan McNab of Dundurn Castle, in Hamilton, Ontario, incorporated into his domestic environment a scullery, pantry, coolroom, ice pit, brewery, butler's plate pantry, servants' quar-

*Top: An unusual apple peeler. Above: Coffee and all the equipment necessary to enjoy it were among the many commodities imported into the colonies.*

# Hearth and Home

*Interpreters at Kings Landing play the role of a family in their dining room about to enjoy soup as part of their early-afternoon dinner.*

ters, and a laundry room. Hot, dim, lit by gaslight and candlelight, the kitchen at the castle is partially underground. The scullery girl had her own fireplace to heat the dishwater, while the cook and her assistant had both a roasting hearth and an iron range. Some large, refined houses had iron hob grates, rare in Canada, or brick hot hearths (potagers) to supplement their fireplaces and iron stoves. Such cooks had an army of specialized culinary utensils: a ham kettle, a fish fryer, an omelette pan; potato beetles, sugar nippers, nutmeg graters, lemon reamers, cherry pitters; wooden rolling pins, brass skimmers, tin patty pans, bone spoons, copper bowls, silver platters, steel knives, pewter pudding moulds; and so many other possibilities. Great numbers of platters and tureens covered by large domes kept food hot as it travelled from kitchen to dining room, and they were arranged in regimented rows on wall hooks or bureau shelves when not in use. The finer the house, the more complex the culinary spaces and the more varied the culinary equipment—and the finer the bills of fare and their presentation were expected to be in the formal dining room. The

# City Cooks and Elegant Fare

*Though hired cooks often worked in smoky, uncomfortable workplaces, their professional reputation depended on their "exquisite sensibility." These civilian cooks at Fort William check their work in progress.*

cook's or chef's reputation was at stake, and therefore his or her employer's.

A professional reputation was built on experience and artistry, important in understanding and interpreting cooking instructions that called for fires that were "brisk and sharp," "gentle," "hot and clear," and similar imprecise, descriptive phrases. Adjusting heat levels required either adding another log to increase the temperature or moving the pots to hotter or cooler places by shifting them back and forth along the crane or up and down on hooks, twisting them from front to back, or manoeuvring the crane in and out. A good hearth cook, in tune with her senses, constantly moved around her fire, stepping back momentarily to reduce its intensity against her skin, then moving forward to

*Baked goods could be kept protected from flies inside a pie safe, commonly found in nineteenth-century kitchens and pantries.*

*A cook must be quick and strong of sight: her hearing most acute, that she may be sensible when the contents of her vessels bubble, although they be closely covered, and that she may be alarmed before the pot boils over; her auditory nerve ought to discriminate (when several saucepans are in operation at the same time) the simmering of one, the ebullition of another, and the full-toned warbling of a third. It is imperiously requisite that her organ of smell ... may distinguish the perfection of aromatic ingredients ... above all her olfactories must be tremblingly alive to mustiness.... It is from the exquisite sensibility of her palate, that we admire and judge of the cook....*

All this in a workplace that was often a fiercely uncomfortable area, for lady, housewife and hired help alike. Cooks worked in hot environments, breathed in too much smoke, and risked injury from flames and sparks. Kitchens were noisier, smellier, darker, dirtier, and more dangerous, then, despite the cook's or employer's best intentions. Kitchens, even the finest, were filled with the clanging of brass mortar and pestle, the ever-present crackle of flames, mice in the pantry, shadowy corners too far from a lamp, the smells of dog and freshly scraped nutmeg, the steady plop of raindrops from a roof leak, the chance of a smouldering log rolling onto the floor. And the everlasting annoyance of flies, which cooks tried to keep off their bowls of batter

smell, taste, listen to, and manipulate her pots instinctively. Robert Roberts, a butler and the author of the best-selling *House Servant's Directory: A Monitor for Private Families*, the first commercially published book written by an African-American in the United States, summarized a cook's sensitivity in 1827:

*Hearth tools, including a salt box, bellows and a goose wing for brushing embers off pot lids and clothing.*

and finished creations using cloths and pieces of paper; pie safes protected baked goods from flies, and dust too. The daily aroma of smoke was intensified if prevailing winds blew the smoke back down the chimney or overcast days slowed natural air circulation. Ash left faint deposits, so regular sweeping and damp dusting was a must. Hearth cooks scorched their fingers and chapped their faces from continual exposure to the fire's intense heat, heaved enormous kettles of scalding water, got varicose veins from standing too much and backaches from lifting heavy pots off the crane.

They also faced the risk of sparks landing on their dresses and aprons, or having their hems carelessly flick the fire. Little burn holes were not uncommon, but sparks generally sputtered out on workaday clothes, since Canadian colonists wore mostly homespun wool or linen cloths, which have a natural suppressant quality. Imported lightweight muslins and cottons for summer were more vulnerable. Sometimes, practical women tucked their dress hems into their apron waistbands to keep them out of the way when cooking, while others preferred ankle-length hems. A goose wing for brushing embers off

# Hearth and Home

*Early cooking stoves were often placed inside or in front of existing fireplaces in order to use the same chimney.*

could begin to boast of indoor plumbing, flush toilets and running water, stoves in each room (even indoor central heating), and gas lighting. Kitchens started to see changes too. An early technological innovation was the iron cooking stove, first available in Eastern Canada in the late 1820s, but only decades later in the Western Interior, where fireplace and bake-oven cooking continued. In the third quarter of the century, the cookstove's adoption coincided with other social shifts, such as more children attending school regularly, the railroad's arrival and the creation of Eaton's and Simpson's mail-order catalogues, from which homemakers and everyone else could order practically anything—from a prefabricated house to linoleum for a kitchen floor to any size cooking pot to any number of brand-name bottled sauces. The availability of a vast array of inexpensive manufactured products led to the diminishment of the variety of heavy and dirty tasks women undertook in their kitchens. While farmwomen still took pride in making their baked goods and preserving their own garden produce, and townswomen had always had the option to go to a neighbourhood baker, confectioner, butcher, fishmonger and grocer to buy ready-made food and condiments, now everyone could choose to purchase tinned fruits and lard, canned salmon, packets of baking powder and granulated sugar, and candles, cloth and cakes. Such products reduced the labour involved in preparing daily meals and supplying the family's daily needs. All these changes combined to slowly but inexorably diminish the household hearth as the main locus of female domestic work.

their skirts, or cleaning ashes off lids, hung from many mantels. Some dresses had detachable lower sleeves, a sensible option in a kitchen. Most women wore some type of white housecap to protect their hair from smoke and keep it clean longer.

In the last half of the nineteenth century, changes in the household started to accelerate. Comfortable middle-class urban households

# 4 Cooking for a Living

One Quebec morning in 1804, Thomas Verchères de Boucherville bought from an old market woman "a small pie that looked superb," but which disappointed him. Perhaps she was a poor widow using one of her skills to eke a meagre liv-

*Top: A tin beer stein. Above: Inns and taverns were often run by couples whose homes became a source of their family income. Here a tavern keeper at Kings Landing offers up some India pale ale.*

# Hearth and Home

*Top: The King's Head Inn at Kings Landing. Above: Until almost the end of the 1800s, women could work in taverns with little social criticism.*

ing by making market foods, such as pies, rissoles, and sausages, in the fireplace in her rented rooms. The inevitable availability of such market and street foods, even within a few years after settlement began, was one sign of the nascent business climate; after all, people needed to eat when away from the home fire too. In the following decades and centuries, many a widow, spinster, farmwife, and family made such articles as pickles, jams, vinegars, ice creams, and cakes in their kitchens at home to make some money by attending a fair or market, or hawking the food from door to door. Some household kitchens doubled as commercial cooking areas. Many inns and taverns, for instance, were run by couples whose homes became a

# Cooking for a Living

temporary or the primary source of familial income.

When Pehr Kalm, the Swedish naturalist, sailed up the St. Lawrence River in summer 1749, he was hosted by English and French riverside settlers, since no inns yet existed. However, in towns, and increasingly beyond them, colonial Canadians could patronize a variety of places to eat outside the home—chop houses, private clubs, cafés, inns and auberges, barrooms and taverns, market stalls, tea gardens, and coffee shops—and could hire a caterer to provide specialties and delicacies for dinner parties and balls. A distinction was made between women who scratched a living by cooking and professional cooks (women) and chefs (men).

*Tavern and hotel guests often ate at a communal table, helping themselves and each other to the dishes in front of them.*

### TAVERN AND INN COOKS

Taverns and inns provided opportunities to fill up quickly while the coach horses were changed, to eat breakfast after staying overnight during a journey, and to enjoy communal dinners. Seeing the smoke curling skyward from a chimney glimpsed through the dense forest must have uplifted many a weary traveller. Those who stopped for sustenance recorded their reactions to the cooking, and not surprisingly the food was described as everything from awful to delicious.

> Our inns are bad … the proverb of "God sending meat and the Devil cooks" never was so fully illustrated as in this country; for with the superabundance of raw material, the manufactured article of a good dinner is hardly to be found in a public-house.

Such hyperbolic accounts of bad accommodations and worse food were balanced by good ones:

> Our bustling hostess . . . quickly prepared us an excellent dinner, consisting of warm fowls (which, by-the-bye [sic], we shot for her), warm bread, butter as yellow as butter-cups, eggs with savoury ham, and everything betokening the Land of Plenty.

Early inns and taverns were often rude log houses, where family and visitors ate together of the basic meat-and-potatoes fare, but

## Hearth and Home

*Roadside inns commonly served quickly prepared sausages and onions.*

through the nineteenth century, hotel dining rooms became more sophisticated, employing professionals to make an expanding variety of fashionable soups, entrées, moulded creams, and sauces in the French style. By century end, hotel dining was the epitome of elegance.

Just as it did in private households, much of the action in early inns and taverns centred around the cooking fireplace (and later cookstove)—everything from heating ale for the hot toddies requested by shivering stagecoach travellers to melting the cheese for Welsh Rarebit to roasting and grilling the plentiful meats for which roadside inns were infamous. Sausages and onions, grilled beefsteaks, pork and mutton chops were common food for travellers, one reason being the speed with which they could be prepared—just a few minutes on a grill. In Nova Scotia and Montreal alike, English inns and coffee houses advertised "hot mutton pies" and "excellent beef soup." Good rural inns had gardens and orchards to supply the kitchen, just like a private household. On their way to Niagara Falls for a visit, Mary O'Brien recorded this incident:

> *We stopped for the night at what was once a very good inn. It is just now dismantled by a recent distress, which circumstances may also account for the ill-humour for the barmaid. She well nigh threw the viands at our heads because one of the company complained that the toast was burned. All of them laughed to see her make a slop basin of the fireplace.*

The next morning, "Breakfast . . . was served for us by two Yankee girls."

Mrs. Montgomery's fireplace, still seen today at Montgomery's Inn Museum in Toronto, was the focal point of a large kitchen that served both the barroom and the travellers' dining room. She had only two windows, although extra light came from the window in her pantry. Her back door opened onto a yard, where she found her stacks of firewood, her well, her chicken pen, her kitchen garden, and her orchard. Tavern mistresses did not leave their work behind at the end of the workday, because travellers showed up at all hours of day and night. Even when she retreated into a private space (if she had one), each tavern cook was still on call. One landlady-cook threw a temper tantrum when the Moodies appeared in her kitchen for breakfast at "an unreasonable hour," but the daughter Almira, "to do [her] justice … prepared from her scanty materials a very substantial breakfast in an incredibly short time." Only in the last third of the nineteenth century did it come to be considered indecent for women to operate inns, cook in tavern kitchens and serve in barrooms.

**CONFECTIONERS AND CATERERS**
Celebrating special events required special foods. If one could afford it, even as early as the 1680s in Montreal and Quebec City, a professional caterer (*traiteur*), pastry cook (*pâtissier*), or confectioner (*confiseur*) could supply for parties and dances a wide array of beautifully decorated cakes, sugar sculptures, roasted meats, brioches, julienne soups, comfit candies, "superior pickles" or elaborate meat pies that supplemented those made by the household cook or chef. One *traiteur* advertised that "those who feel it is appropriate to have large or small meals at home" could purchase from him "feasts and meals at a reasonable price and also . . . all sorts of pastries and iced cheeses," including cream ices, blancmanges, and mousses. As independent businessmen-cooks, often with extensive training in the culinary arts, they ran their businesses out of their home kitchens or rented premises, sometimes with employees and apprentices, including unremunerated wives and offspring, to whom they passed their culi-

*A tavern mistress serving some tasty stew and bread.*

nary secrets. Expensive cookbooks for professional chefs were sold in New France, such as La Varenne's *Le cuisinier françois* (1651) and Liger's *La nouvelle maison rustique* (1755).

Professional male French chefs, confectioners, and their assistants who worked for the governor, the intendant, and the bishop, had lavishly equipped kitchens and pantries provisioned by large vegetable gardens and orchards, plus wine cellars. Often, we know their names. Chef Duval, for instance, who was trained in La Varenne's style of early classic French cuisine, was employed by Governor Duquesnel of Louisbourg in the 1730s. Chef

*A large dish pantry at the Fortress of Louisbourg. Hired cooks often worked in kitchens more elaborately equipped than their own.*

Petit, "so distinguished a French artist," was on Governor Craig's staff in Quebec City before he opened a catering business from which discerning housewives could purchase breads, cakes, pastries, and sugar centrepieces for their dining tables.

Following Parisian precedent, the first café in the colony was opened in 1739 in the Lower Town in Quebec. Canada's first true restaurant is considered to be Charles-René Langlois's *Hôtel de la Nouvelle Constitution* in Quebec City, which opened and closed in 1792. Canada's first private club had also been short-lived two centuries earlier, when, to promote good humour and health during the bitter winter of 1606–07, Samuel de Champlain's men formed a banqueting club called *l'ordre du bon temps* (The Order of Good Cheer) at the Habitation of Port Royal. They "vied with each other to see who could do the best" hunting and catering.

The first baker to open shop in York (now Toronto) was a John Horton, whose advertisement in the *Upper Canada Gazette* of August 30, 1800

> *begs leave to inform his Friends and Public, that he carries on the Bakeing [sic] business at his Bake-house . . . ; where he keeps Bread and Cakes for sale on the most reasonable terms. Also some excellent Smoked-beef.*

Others soon challenged him, so that, in 1834, the year York was renamed Toronto, seventeen bakers, sausage makers, breadmakers, and confectioners were in business, including Mrs. Lumsden, the gingerbread baker, whose husband was a tailor. Franco Rossi, an Italian confectioner, was probably the first to sell ice cream in York, while Lefébvre, a Parisian ice-cream maker, started in Quebec about 1820. Other Italian immigrants started food enterprises and restaurants too, competing with cooks who were self-proclaimed French *artistes*.

### Feeding a Fortful

In the initial settlement decades before soldiers' housing was built, both the French and later the British armies billeted soldiers and officers in inns and private homes, where they partook of the families' meals in their big communal rooms and perhaps helped out by

## Cooking for a Living

*Soldiers commonly cooked for their mess groups in cookhouses, such as this one at Fort Henry. Notice the "stew holes" along the side wall.*

chopping firewood and hauling water. After moving into garrison, the soldiers cooked for themselves in the cookhouse, combining rations from the commissariat and whatever they could purchase, hunt, fish, steal, grow, or barter. British soldiers were grouped into messes of eight or ten men, each one of whom took turns cooking for their group—as well as the wives and children of the married soldiers—either in the cookhouse or on the campfire when in camp. The equipment was limited: "We have an iron pot which serves for teapot, roaster and boiler," wrote a soldier. Armies relied on imported salt pork, rice, and flour until local farmers could supply enough wheat, rye, pork, and beef. Soldiers grew tired of the inevitable salt pork and pease porridge. Fresh bread and hardtack came from the regimental baker—sometimes a professional, sometimes a hapless corporal assigned the duty. Out West, Indian and Métis women made pemmican for employees of the Hudson's Bay Company, the North West Mounted Police, and the Canadian army.

Social inequality between soldiers and officers meant they ate and drank differently. Officers, generally being able to afford servants, superior food, and sometimes regimental plate, dined better than the average soldier, expecting meals that blended traditional (roast beef with gravy, suet pudding) and fashionable (spinach soup puréed in a tammy) cooking.

# Hearth and Home

*Civilian bakers were often contracted to provide daily bread for both the soldiers and officers.*

As they assembled in the mess dining room in their formal attire, prodigious quantities of wine and spirits were imbibed along with the fine food. A number of British-Canadian officers' kitchens had British-style cooking grates, rather than fireplaces or stoves. A surviving example is at Stanley Barracks in Toronto.

Little is known about British officers' cooks, other than that they were often soldiers' wives and daughters along with local civilians, employed by a mess steward, who operated the officers' establishment rather like an institutional catering service. Since their sole aim was to produce large daily dinners for the officers and their guests, the kitchens were hot and busy places, and perhaps even chaotic.

Complaints were often lodged against these cooks, who were not necessarily trained. Gentlewoman Hannah Jarvis lamented that "soldiers' Wifes are all we can get." Nonetheless, some officers ate very well indeed, and some regiments were known for hosting splendid dances for the townspeople.

A cook named Mrs. Chapman survived the Battle of York on April 13, 1813, according to an eyewitness account:

*I . . . perceived we were to leave the Garrison, And I went . . . to our Quarters got my Coat, advised Mrs. Chapman a Woman that Cooked for us to come away & and as … [we] returned out at the Gate the Magazine blew up & for a few Minutes . . . [we were] in a Horrid-situation, the stone falling thick as Hail & large one's sinking into the very earth.*

Apparently, she did not abandon the officers' mess kitchen when the battle began, but only once the Americans started to advance. Perhaps she continued to cook dinner, thinking "her" officers would want a meal after all the fuss.

# Epilogue: From Wood Fire to Coal Fire

For countless millennia before our era of gas, electricity, and microwaves, dinner was cooked over a fire. Today we still love to gather around a living-room fire on a cold winter evening, or a beach bonfire on a starry summer night, or a campfire after a hike or canoe trip. A fire in a hearth still represents community and the sharing of food. Part of the fun of camping is making three meals a day over the little fire, even if there's a kerosene stove along too. Marshmallows and wieners twirled on sticks over beach bonfires continue to entice us—and it is a rare suburban backyard without a barbecue. Simultaneously, interest in the art of wood-fired ovens has revived in recent years, as shown by the

*Top: A hearth shovel. Above: The bright, hot flames inside the fireplace contrast with the closed heat of the cookstove standing in front of the boarded fireplace. Hearth cooking was dirtier and hotter than using a cookstove.*

49

*Eventually, kitchen fireplaces were boarded up in favour of stoves, and new houses were sold with new stoves already installed.*

number of upscale pizzerias which use them and the books on building instructions. You can even roast a chicken in a tin kitchen in front of your living-room fire or have a wood-fire oven installed in your household kitchen.

The adoption of the cookstove in the second half of the Victorian era caused a decline in the skills particular to open-hearth and brick-oven cooking. When cookstoves started to take precedence over traditional hearths, cooks had to adjust their skills to suit the new-fangled contraptions. The age-old instinctive dance, in which the crane was continuously shifted and the embers shovelled onto the hearth for the long-handled and three-legged pots, became unnecessary. Instead, cooks learned to coax the great beasts by adjusting various dampers, levers and grates to control heat circulation and avoid smoke backfilling the house. Wood, especially in urban areas, was replaced by cheaper and plentiful coal.

In the last quarter of the nineteenth century, though the kitchen table lit by a kerosene lamp and warmed by the cookstove became the family focus, people remained nostalgic for the kitchen hearths:

> *In [the kitchen] was the large fire-place, around which gathered in winter time bright and happy faces; where the old men smoked their pipes in peaceful reverie, or delighted us with stories of other days; where mother darned her socks, and father mended our boots; where the girls were sewing, and uncles were scraping axe-handles with bits of glass to make them smooth.*

Visitors to historic sites with lively kitchen activity continue to be enthralled by the smells, sounds, and tastes of hearth cooking. The study of hearth cookery has flourished in recent years, because the best way to understand it is to re-enact it. The cook at her hearth is an enduring image of the past.

# Bills of Fare

Visitors who watch open-hearth cooking interpretation at historic sites, or who try it themselves, are often surprised, even impressed, by the sophistication and variety of the meals and dishes that cooks mustered, and delighted by the rich flavours yielded. Here is a taste, in words, of the sorts of meals prepared over fire in early Canada.

## Christmas and New Year's Day

From Joseph Willcock's Diary, Library and Archives Canada, MG 24 C1. Quoted in Edith Firth, *The Town of York, 1793–1815 A Collection of Documents of early Toronto*, p. 232.

Joseph Willcock was a prolific diarist, journalist and anti-authoritarian politician in early Ontario.

"Thursday, December 25, 1800. Went to Church Weekes [a friend] dined with us we had for dinner soup roast beef boiled Pork Turkey Plumb Pudding & minced pies We had a supper for the first time in my remembrance. I came to bed at 12 It was a very fine day Playter called for some camomile

"Thursday, January 1, 1801. Miss Russell & I went out in the sled ... I upset the sled coming home McGill the two Ridouts the Solicitor Gen'l & Ruggles called to pay their respects We had for dinner Boiled beef, a roast Pig & minced meats. ... It froze the bay across I went to bed at 10."

## British Soldier's Rations

From RGB, microfilm, C3502, vol. 1168, pp. 112 and 105, Library and Archives Canada. Quoted in "Eating on the March," Richard Feltoe, *Consuming Passions: Eating and Drinking Traditions in Ontario*, pp. 17-18.

**Ration Scale for Troops in Garrison
19 March 1812
Per Man Per Day**

1 lb. Flour
9 ½ oz. Pork
3/7 pint Pease
1 ½ oz. Rice

⁶⁄₇ oz. Butter or 1 ³⁄₇ oz. Pork Fat

**Ration Scale for Troops in Garrison**
**15 July 1813**
**Per Man Per Day**

1 lb. Flour or Biscuit
²⁄₇ Pint Pease
9 ½ oz. Beef Flour and ½ oz. Beef
¾ Pint Oatmeal
⁶⁄₇ oz. Sugar
⁶⁄₇ oz. Butter
1 ⁵⁄₇ oz. Cheese, Rice or Cocoa
2 lbs. Beef (twice a week)
1 lb. Pork (twice a week)

Women received one-half, and children one-third, of men's rations.

## Breakfast

From *The Dalhousie Journals*, vol. 1, p.87 edited by Marjory Whitelaw.

George Ramsay, 9th Earl of Dalhousie, was Lieutenant-Governor of Nova Scotia 1816–1820. While travelling around Nova Scotia, he received meals from many people, although he seldom described them. This passage concerns a meal received in the Parrsborough district from an old settler couple who were Loyalists from New York.

"10ᵀᴴ July [1816]. We got an excellent breakfast of tea & coffee, but the sugar was the maple & has an aromatic taste, the bread was very good, the produce of the farm—with variety of sweetmeats & common fruit preserves made by themselves."

## Private Sleigh Ride Dinner

From Quebec Driving Club, Library and Archives Canada, C.84669. Quoted in Lafrance and Desloges, *A Taste of History: The Origins of Québec's Gastronomy*, p. 87.

This two-course bill of fare for dinner after a sleigh ride was eaten by twenty members of the Quebec Driving Club in Quebec City, some time in winter 1831. In the first course, the soups were consumed first, followed by the roasts and accompanying vegetables, then the meat pies and cutlets. This is how it was laid out on the menu card.

**1st Course**

| | |
|---|---|
| Hare | Ham |
| Soups: Giblet | Tongues |
| Gravy | Venison Pies |
| Haunch of Venison | Roast Fowl |
| Roast Beef | Fricasseed Fowl |
| Round of Beef | Hashed Hare |
| Roast Turkey | Oyster Patties |
| Raised Pies | Mutton Cutlets |
| Pork Cutlets | Beef Steak Pie |
| Curry | |

**2nd Course**

| | |
|---|---|
| Jelly | Coffee Cream |
| Punch Jelly | Blancmange |

*Bills of Fare*

Mince Pies       Stewed Apples
Italian Cream    Tarts

## Meals on the Surveying Trail

From *Jukes' Excursions: Being a revised edition of Joseph Beete Jukes' "Excursions In and About Newfoundland During the Years 1839 and 1840,"* edited by Robert Cuff and Derek Wilton, pp. 49, 50, 59, 64.

Joseph Beete Jukes was a geological surveyor who hiked, boated and rode through much of Newfoundland in 1839 and 1840. In between visiting isolated communities, he and his crew often travelled for days without seeing any settlers or Natives.

"May 21st. The bake-pot was put in requisition this morning for our service, and a lot of little cakes made by our hostess for breakfast, at which she gave us also some fresh herrings.

"August 14th. These [a goose, a gull and a loon], with the addition of a lump of salt pork, we boiled all together in our boat's kettle, and, thickening the broth with a little flour, we generally cleared off all its contents, both fluid and solid. To the bouillée of the kettle we added some molasses-tea and a couple of common sea-biscuits both morning and evening, taking a biscuit or so during the day as we had leisure or appetite.
"August 30th. We had a grand soup-making this evening of black ducks, which are excellent eating.

"September 3rd. We waited in vain ... for more [caribou], but luckily shot a couple of divers [loons], which served us for supper. The men's bread was now all gone, and we had nothing but tea and our guns for support. I had, however, kept in my knapsack a private store of half a dozen biscuits and a piece of ham, against an emergency, as also a bunch or two of raisins, and we still had a little flour left in the bag, for thickening our soup."

## House-Raising Bee on Colonial Frontier

From Frances Stewart, 1841, from *Our Forest Home: Being Extracts from the Correspondence of the Late Frances Stewart*, pp. 173-176.

One of the most anticipated events was the gathering associated with the communal raising of a barn or a house. Pioneer housewives, with the help of their daughters and female friends, undertook prodigious culinary labours to feel all the hungry men who participated. When Thomas and Frances Stewart erected their second house on July 9 and 10, 1841, Frances oversaw the serving of five meals to between fifteen and thirty people; for three of these meals she recorded the bills of fare.

DINNER
"Dinner was called at half-past twelve." Dinner was the Canadian farmer's word for

the midday meal. "We had a famous dinner, substantial though not very elegant … [W]e had a roast pig and a boiled leg of mutton, a dish of fish, a large cold mutton pie, cold ham and cold roast mutton, mashed potatoes and beans and carrots; … For second course we had a large rice pudding, a large bread and butter pudding, and currant and gooseberry tarts."

## TEA

"It had settled into a wet evening." Tea was the early evening meal of cakes and tea, not as substantial as dinner, which designated the end of the afternoon. "The gentlemen drank the punch for which they would not delay after dinner, and those who liked smoked cigars. … We had a large tea table; a tray and a tea-pot at either end were presided over by Miss Haycock and Anna and there were plenty of cakes, bread and butter and strawberry jam."

## SUPPER

Supper was served at "eleven o'clock" after "we passed the time [since tea] between this amusement [dancing] and singing." Supper was a cold collation served at the end of a ball, or in this case, a country evening dance. "Anna and I had laid the supper table *in the kitchen* which had been prepared for us before the servants had gone to bed. We had a cold ham at one end of the table, a pair of [cold] roast fowls at the other, the intervening space being filled with tongue, cold mutton, cakes, tarts, cups of custard, a few decanters of currant cordial (home-made). Altogether it looked very respectable …".

## French-Canadian Dinner in the New Brunswick Backwoods

From Frances Beavan, from *Life in the Backwoods of New Brunswick*, 1845, p. 33–34.

Francis Beavan was an Irish-born woman who lived on a New Brunswick farm for seven years with her husband and children, before they moved to England. She was a teacher, poet and novelist.

"The present meal consists of fine trout from the adjoining stream, potatoes white as snowballs, some fried ham, and young French beans, which grow here in the greatest luxuriance. We have also a bowl of salad, and home-made vinegar prepared from maple sap, a large hot cake, made with Indian meal, and milk and dried blue-berries, and excellent substitute for currants. Buscuits [sic], of snow white Tenessee [sic] flour, raised with cream and sal-a-ratus. This last article, which is used in the place of yeast, or eggs, in compounding light cakes, can also be made at home from ley of wood ashes, but it is mostly bought in town. A raspberry pie, and a splendid dish of strawberries and cream, with tea (the inseparable beverage of every meal in New Brunswick), forms our repast; and such would it be in ninety-nine houses out of a hundred of the class I am describing."

# Bills of Fare

## Prairie Party

From Charles Mair, 27 November 1868, *Toronto Globe*. Quoted in W.J. Healey, *Women of Red River: Being a Book Written from the Recollections of Women Surviving from the Red River Era*, p. 129.

"We had a very pleasant stay at Fort Garry [Manitoba], and received all sorts of entertainment. They live like princes there. Just fancy what we had at a dinner party! Oyster soup, white fish, roast beef, roast prairie chicken, green peas, tomatoes stewed, stewed gooseberries, plum pudding, blanc mange, raisins, nuts of all kinds, coffee, port and sherry, brandy punch and cigars, concluding with whist until four o'clock a.m. There is a dinner for you, in the heart of the continent, with Indian skin lodges within a stone throw!"

## Western Resourcefulness

From Susan Allison, *A Pioneer Woman in British Columbia*, 1976. Quoted in Beulah Barss, *Come n' Get It at the Ranch House*, p. 22–23.

Susan Allison was the first white woman to settle in Similkameen Valley, British Columbia. Her stove melted in a house fire, but with her children's help she replaced it with a campfire and bake oven.

"It was hard work cooking without a stove. I thought that with all hands helping I could build a bake oven. This I did on the banks of a river using cobblestones and clay. All hands gathered the stones, Will and Beatrice mixed mud, Lily packed it to me and I built it. Someone found a half melted door from the old stove which made a good door for the new oven.

"One day near June I had put three ducks nicely dressed, and a huge custard into the oven, and all was cooked ready to be eaten by hungry people. I had no tea or coffee so browned some dried peas and used them for coffee, which was also ready. Two well-dressed men rode up and said they were starving so I gave them something to eat …. I dished up the dinner and told them to sit down. I thought there would be enough for us but no, they ate all the ducks and pudding. They talked French thinking I couldn't understand them. They said my bread was like cake and the ducks the best they had ever tasted. I felt proud of the oven after that."

Mrs. Carl Diggle, Prince Albert, 1870. Quoted in Beulah Barss, *The Pioneer Cook: An Historical View of Canadian Prairie Food*, 1980, p. 28.

"The people in those days lived mostly on the fresh Buffalo meat in the winter and pemmican in the summer. There would be weeks at a time without tasting bread. Father made a trip to Fort Carleton, sixty miles distant to get some blacksmith work done. Being a friend of the Chief Factor, he was allowed a couple of sacks of flour at $15.00 a sack. He had a little coffee mill in which he often ground enough wheat to make pancakes."

# Sites of Interest

## Historic Sites with Open-Hearth Foodways Programs

Today, a fireplace symbolizes a home, but not long ago its role was more than symbolic. Knowing this, many historic sites in Canada enliven their kitchens with period cooking demonstrations because hands-on experience evokes more understanding than reading cookbooks or historical accounts. Occasionally, visitors are invited to participate by pounding corn in a mortar, flipping a griddle cake, churning butter, or attending an historical cooking class.

### New Brunswick

**Kings Landing Historical Settlement, 1820s to 1909**
20 Kings Landing Road,
Kings Landing, NB E6K 3W3
506 363-4999
www.kingslanding.nb.ca/
This outdoor museum recreates rural Loyalist New Brunswick in the nineteenth century. Several houses have working hearths, ovens and cookstoves.

**Village Historique Acadien, 1770 to 1939**
C.P. 5626
Bertrand, near Caraquet, NB E1W 1B7
506 726-26001 or 1 877 721-2200
www.villagehistoriqueacadien.com/main.htm

**Acadian Odyssey National Historic Site / Monument Lefebvre National Historic Site of Canada**
480, rue Centrale,
Memramcook, NB E4K 3S6
506 758-9808; out of season: 506 536-0720
www.pc.gc.ca/lhn-nhs/nb/lefebvre

### Nova Scotia

**Fortress of Louisbourg, National Historic Site of Canada, early to mid-eighteenth century**
259 Park Service Road,
Louisbourg, NS B1C 2L2
902 733-2280
www.pc.gc.ca/lhn-nhs/ns/louisbourg
Wonderful site with good hearth cooking demonstrations.

## Sites of Interest

### Ross Farm, mid-nineteenth century
4568 Highway 12, RR#2,
New Ross, NS B0J 2M0
902 689-2210
rossfarm.museum.gov.ns.ca
A charming historic farm located on a few acres of the original site. A little bit of open hearth baking is done at Rose Bank Cottage, where members of the Ross Family lived from 1817 until 1969.

## Prince Edward Island

### Le Village de l'Acadie, early nineteenth century
Mont-Carmel, PE C0B 2E0
902 854-2227; toll free: 1 800 567-3228
www.teleco.org/village/contate.htm

## Quebec

### Fort Chambly National Historic Site of Canada
2 De Richelieu Street,
Chambly, QC J3L 2B9
450 658-1585; toll free: 1 888 773-8888; TTY: 1 866 558-2950
www.pc.gc.ca/ihn-nhs/qc/fortchambly
A few special events include open-hearth cookery.

## Ontario – Toronto

### Black Creek Pioneer Village, Downsview, Toronto, 1860s
1000 Murray Ross Parkway,
Toronto, ON M3J 2P3
416 736-1733 general information
www.trca.on.ca/parks_and_culture/places_to_visit/black_creek
This dynamic heritage site depicts rural Ontario of the 1860s. Its forty buildings are centred round the Stong Farmstead (a log cabin dating to 1816, a second house of 1832) on their original sites. Several houses have working hearths and iron stoves. Bread is baked daily in the brick oven at the Halfway House (hotel) of 1849.

The following sites are managed by the Culture Division of the City of Toronto, as are Mackenzie House Museum and Spadina Museum in the Appendix II list.

### Colborne Lodge, 1837
South end of High Park
c/o Spadina Museum, 285 Spadina Road,
Toronto, ON M5R 2V5
416 392-6916
www.toronto.ca/culture/colborne.htm
Built by John and Jemima Howard, the cosy winter kitchen features a small hearth and brick oven, while the summer kitchen has the Howard's iron stove. The Lodge does cooking classes for families several times a year.

### Fort York, War of 1812
100 Garrison Road,
Toronto, ON M5V 3K9
416 392-6907
www.toronto.ca/culture/fort_york.htm
Fort York includes two working kitchens and the archaeological remains of Toronto's oldest extant kitchen (1815) in the 1826 Officers' Brick Barracks. The very active open-hearth cookery program reflects the officers' middle class status, while the knowledgeable male and female Volunteer Historic Cooks represent the soldiers and civilians who cooked for them. One of their spectacular annual fundraisers is the Georgian Dinner, which recreates an officers' banquet of the War of 1812 time period.

### Gibson House Museum, Willowdale, 1851
5172 Yonge Street,
Toronto, ON M2N 5P6
416 395-7432

www.toronto.ca/culture/gibson_house.htm
Erected by the Scottish Gibsons in 1851, this historic site depicts rural life in the mid nineteenth century, despite the urban surroundings. The museum is well known for regularly offering hands-on cooking classes at the hearth and throwing a true Hogmanay, complete with their own haggis.

### Scarborough Historical Museum
1007 Brimley Road,
Toronto, ON M1P 3E8
416 338-8807; event hotline: 416 338-3888
www.toronto.ca/culture/scarborough_historical.htm
Set within the pretty Thomson Memorial Park, the McCowan Log House represents the area's early nineteenth-century Scottish settlers. Their best known special event around period food is Christmas Desserts by Lamplight, featuring a huge variety of forgotten nineteenth-century puddings.

### Montgomery's Inn Museum, 1830-47
4709 Dundas Street West,
Toronto, ON M9A 1A8
phone: 416 394-8113;
event hotline: 416 338-3888
www.toronto.ca/culture/montgomerys_inn.htm
Although built in 1830, by Thomas and Margaret Montgomery, the Inn is restored to its heyday of the late 1840s. The especially large kitchen is animated by particularly well-informed staff and volunteers, mostly at special events.

### Todmorden Mills: Heritage Museum & Arts Centre
67 Pottery Road,
Toronto, ON M4K 2B8
416 396-2819; event hotline: 416 338-3888
http://www.toronto.ca/culture/todmorden_mills.htm
The Mills played a vital role in the early food and beverage industries. Today, they specialize in nature programing, although for several years now they have offered a series of children's cooking classes cleverly dubbed The Cast Iron Chef.

## Ontario – outside Toronto

### Battlefield House, early nineteenth century
77 King Street West, P.O. Box 66561,
Stoney Creek, ON L8G 5E5
905 662-8458
www.battlefieldhouse.ca
The hearth is used a lot in educational programs and special events.

### Dundurn Castle National Historic Site, 1850s
610 York Boulevard,
Hamilton, ON L8R 5H1
905 546-2872
www.myhamilton.ca. Follow the links through Culture.
Within the 1850s domestic offices in the basement, the Castle has a fabulous working kitchen with knowledgeable historic cooks dressed in costume. They offer many hands-on cooking classes and several special events based on historic foods.

### Fort George, War of 1812
Queen's Parade,
Niagara-on-the-Lake, ON L0S 1J0
905 468-4257
www.niagara.com/~parkscan. Follow the links to Fort George.

# Sites of Interest

**Fort William Historical Park, 1803-1821**
1350 King Road,
Thunder Bay, ON P7K 1L7
807 473-2333; TTY: 807 473-2307
www.fwhp.ca

**Hutchison House Museum, c 1840-60**
270 Brock Street,
Peterborough, ON K9H 2P9
705 743-9710 or 1 866 743-9710
www.nexicom.net/~history/index.html
The large hearth in the keeping room is often active during special events and educational programs.

**Ireland House at Oakridge Farm, 1835**
2168 Guelph Line,
Burlington, ON L7P 5A8
905 332-9888 or 1 800 374-2099
www.geocities.com/EnchantedForest/Cottage/6640/IHMuseum/IHMuseum.html

**Lang Pioneer Village, nineteenth century**
104 Lang Road,
Keene, ON
c/o 470 Water Street,
Peterborough, ON K9H 3M3
705 295-6694 or 1 866 289-5264
www.langpioneervillage.ca
Thursdays are "From the Hearth" day.

**Ste. Marie among the Hurons, Midland, 1640s**
Highway 12 East,
Midland, ON L4R 4K8
705 526-7838
www.saintemarieamongthehurons.on.ca
Standing recreated on its original foundations, this seventeenth-century wood fortress was headquarters for the French Jesuit mission to the Wendat (Huron) nation. The hearths represent the European and Wendat technologies.

**Pickering Museum Village**
3 kilometres east of Brock Road on Highway #7 (west of Westney Road)
c/o The City of Pickering, One The Esplanade,
Pickering, ON L1V 6K7
905 683-8401
www.cityofpickering.com/museum

**Upper Canada Village, 1860s**
13740 County Road 2,
Morrisburg, ON K0C 1X0
1 800 437-2233 or locally 613 543-4328
www.uppercanadavillage.com/home.htm
The bread made fresh daily in the brick oven at the bakery is for sale. Several of the houses have working hearths.

**Westfield Heritage Village, nineteenth century**
1049 Kirkwall Road,
Rockton, ON L0R 1X0
519 621-8851 or 1 800 883-0104
www.westfieldheritage.ca

## Manitoba

**Lower Fort Garry National Historic Site, 1850s**
5925 Highway 9,
St. Andrews, MB R1A 4A8
1 888 773-8888; TTY: 1 866 787-6221
parkscanada.pch.gc.ca/lhn-nhs/mb/fortgarry

## Alberta

**Ukrainian Cultural Heritage Village**
8820 – 112 Street,
Edmonton, AB T6G 2P8
780 662-3640
www.cd.gov.ab.ca/uchv
This lively recreated village features several Ukrainian farmsteads dating from 1892 to 1930 from east central Alberta. They cook and bake in their three historic pich (clay ovens), and

demonstrate the making of sauerkraut, cabbage rolls and borsch. Occasionally, they offer cooking classes.

### British Columbia

#### Fort Langley National Historic Site of Canada, 1827
PO Box 129, 23433 Mavis Ave,
Fort Langley, BC V1M 2R5
604 513-4777
www.pc.gc.ca/lhn-nhs/bc/langley

## Historic Sites with Cookstove and Gas Stove Programs

#### Doon Heritage Crossroads, 1914
10 Huron Road at Homer Watson Boulevard,
Kitchener, ON N2P 2R7
519 748-1914; TTY: 519 748-0537
www.region.waterloo.on.ca (follow the links through "visiting" to "museums")
Although the site's focus date is 1914, the foodways program uses the open hearth in one building and the cookstove in the others.

#### Louck's Farm House, 1846
Upper Canada Village, 13740 County Road 2,
Morrisburg, ON K0C 1X0
1 800 437-2233 or locally 613 543-4328
www.uppercanadavillage.com/home.htm
Prosperous farmhouse restored to reflect the 1860s.

#### Schneider Haus, c 1816
466 Queen Street South,
Kitchener, ON N2G 1W7
519 575-4491 or 519 742-7752
www.region.waterloo.on.ca (follow the links through "visiting" to "museums.")
The active foodways program concentrates on Mennonite food of the early to mid-nineteenth century.

These two sites are part of the Culture Division of the City of Toronto. See also Appendix I.

#### Mackenzie House Museum, 1860
82 Bond Street,
Toronto, ON M5B 1X2
416 392-6915; event hotline: 416 338-3888
www.toronto.ca/culture/mackenzie_house.htm
The large two-room basement kitchen has an iron range called "Our Favorite," used on weekends by the Volunteer Historic Cooks. The Kitchen Cultures program welcomes immigrant ethnic groups into the 1850s kitchen to prepare their foods for visitors to taste.

#### Spadina Museum: Historic House and Gardens, early twentieth century
285 Spadina Road,
Toronto, ON M5R 2V5
416 392-6910
www.toronto.culture.ca/spadina.htm
The 1898 kitchen features an in situ 1936 gas stove called "Miss Canada" and a 1942 gas refrigerator, both used by an active group of costumed Volunteer Historic Cooks. The basement has the archaeological remains of the 1818 hearth and oven. Many annual special events – such as Edwardian Tea and Strawberry Festival – include

refreshments from cookbooks appropriate to the Austins. Cooking classes are offered occasionally.

**Woodside National Historic Site of Canada, 1890s**
528 Wellington Street North,
Kitchener, ON N2H 5L5
519 571-5684
www.pc.gc.ca/lhn-nhs/on/woodside
Restored to the 1890s, this is the boyhood home of Canada's Prime Minister William Lyon Mackenzie King. Visitors see lots of food preparation on the magnificent cookstove.

# Notes

Below are the sources of the quotations reprinted within the text. The quotations are listed in the order in which they appear. For complete bibliographic information on the sources, see the Selected Bibliography.

## Introduction

*"After a tedious walk"*
   Frances Stewart, *Our Forest Home*, Toronto: Gazette Printing, 1902 ed., p. 75.

*"prosperity depends on female industry"*
   Frances Beavan, *Life in the Backwoods of New Brunswick* [1845], St. Stephen, NB: Print'n Press, 1980, p. 79.

## 1: Pioneer Hearths

*"We find the fireside"*
   Louisa Collins, *Louisa's Diary, The Journal of a Farmer's Daughter, Dartmouth, 1815*, Halifax: Nova Scotia Museum, Nimbus Publishing, 1989, p. 20.

*"fit for the table of the most fastidious epicure"*
   Susanna Moodie, *Roughing It in the Bush; or, Life in Canada* [1852], Toronto: McClelland and Stewart, 1989, p. 302.

*"large, light kitchen"*
   Stewart, *Our Forest Home*, p. 29.

*"on waking"*
   Stewart, p. 39.

*"We first put on a* **back log**"
   Stewart, p. 48.

*"The best wood for fires"*
   *The Maple Leaf, a juvenile monthly*, Montreal, May 1853, vol. 2, no. 5, p. 159.

*"often brought home pine knots"*
   Stewart, p. 51.

# Notes

**"I really think"**
Letitia Hargrave, quoted in Beulah Barss, *The Pioneer Cook*, Calgary, AB: Detselig Enterprises Ltd., 1980, p. 20.

**"first Canadian loaf"**
Moodie, pp. 120–21.

**"I had just finished the first stage"**
Mary O'Brien [Audrey Saunders Millar, ed.], *The Journals of Mary O'Brien, 1828–1838*, Toronto: MacMillan of Canada, 1968, p. 141.

## 2: Prosperous Cooks and Kitchens

**"a large kettle"**
Harriette Cowan, quoted in W. J. Healey, *Women of Red River: Being a Book Written from the Recollections of Women Surviving from the Red River Era*, Winnipeg: The Women's Canadian Club, 1923, p. 38.

**"Jams and jellies"**
Mrs. Carl Diggle, Questionaire Response, Saskatchewan Archives Board, University of Saskatchewan, 1952, quoted in Barss, *The Pioneer Cook*, p. 90.

**"take great Care that the Bag or Cloth be very clean…"**
Hannah Glasse, *The Art of Cookery, made Plain and Easy* [London: 1747, p. 70]; facsimile edition, Totnes, Devon, UK: Prospect Books, 1995.

**"Oxen and sheep were killed"**
Elizabeth Norquay, quoted in Healey, p. 210.

## 3: City Cooks and Elegant Fare

**"dined as good as those in France"**
Jesuit Beschefer, quoted in Marc Lafrance and Yvon Desloges, *A Taste of History: The Origins of Québec's Gastronomy*, Montréal: Les Éditions de la Chenlière, 1989, p. 10.

**"during the dancing"**
*Royal Gazette*, Halifax, 1792, quoted in Marie Nightingale, *Out of Old Nova Scotia Kitchens*, Halifax: Petheric Press, 1970, p. 8.

**"A cook must be quick and strong of sight"**
Robert Roberts, *The House Servant's Directory*, [1827], Old Saybrook, CT: Applewood Books, 1993, p. 176. Roberts was butler to The Hon. Christopher Gore, Governor of Massachusetts. Though few Canadian housewife-cooks are likely to have read Roberts' book, since he directed it to upper-scale servants, he captured in this exuberantly poetic passage the physicality of cooking down-hearth.

## 4: Cooking for a Living

**"a small pie that looked superb"**
Thomas Verchères de Boucherville [1804]

quoted in Lafrance and Desloges, *A Taste of History*, p. 35.

**"Our inns are bad"**
William Dunlop, *Tiger Dunlop's Upper Canada* [1832], quoted in *Consuming Passions: Eating and Drinking Traditions in Ontario*, Willowdale, ON: Ontario Historical Society, 1990, p. 40.

**"Our bustling hostess ..."**
quoted in Edwin Guillet, *Pioneer Inns and Taverns*, Toronto: Ontario Publishing Co., 1958, vol. 4, p. 94.

**"hot mutton pies" and "excellent beef soup"**
Nightingale, *Out of Old Nova Scotia Kitchens*, p. 9.

**"We stopped for the night"**
O'Brien, *Journals of Mary O'Brien*, p. 74.

**"an unreasonable hour"**
Moodie, *Roughing It in the Bush*, pp. 58–59.

**"superior pickles"**
Lafrance and Desloges, p. 63.

**"those who feel it is appropriate"**
Lafrance and Desloges, p. 67.

**"so distinguished a French artist"**
Lafrance and Desloges, p. 40.

**"vied with each other"**
Samuel Champlain, quoted in Jo Marie Powers, "L'Ordre de Bon Temps: Good Cheer as the Answer," Oxford Symposium on Food & Cookery, *Feasting and Fasting: Proceedings*, Harlan Walker, ed., London: Prospect Books, 1990, p. 164.

**"begs leave to inform"**
*Upper Canada Gazette*, 30 August 1800.

**"We have an iron pot"**
Thomas Ridout, *Reminiscences of Niagara* [1813], quoted in *Consuming Passions*, p. 25.

**"soldiers' Wifes are all we can get"**
Hannah Jarvis, quoted in Fiona Lucas, "Officers' Mess Servants at Fort York," in *Explore Historic Toronto*, Issue 7, Fall/Winter 1994, pp. 8, 12.

**"I ... perceived we were to leave the Garrison"**
Ely Playter Diary, 27 April 1813, excerpted in Edith Firth, *The Town of York, 1793–1815: A Collection of Documents of Early Toronto*, Toronto: Champlain Society and University of Toronto Press, 1962, p. 280.

## Epilogue: From Wood Fire to Coal Fire

**"In [the kitchen] was the large fire-place"**
Caniff Haight, *Country Life in Canada* [1885], Belleville, ON: Mika Publishing Co., 1986, p. 10.

# Selected Bibliography

Barss, Beulah. *The Pioneer Cook: An Historical View of Canadian Prairie Food.* Calgary, AB: Detselig Enterprises Ltd., 1980.

Barss, Beulah. *Come 'n Get It at the Ranch House.* Calgary, AB: Rocky Mountain Books, 1996.

Bates, Christina. *Out of Old Ontario Kitchens: A Collection of Traditional Recipes of Ontario and the Stories of the People Who Cooked Them.* Toronto: Pagurian Press, 1978.

Beavan, Frances. *Life in the Backwoods of New Brunswick* [1845], St. Stephen, NB: Print'n Press, 1980.

Bell, John, ed. *Halifax: A Literary Portrait.* Lawrencetown Beach, NS: Pottersfield Press, 1990.

Boily, Lise and Jean-François Blanchette. *The Bread Ovens of Quebec.* Ottawa: National Museums of Canada, 1979.

Carson, Jane. *Colonial Virginia Cookery: Procedures, Equipment, and Ingredients in Colonial Cooking* [1968]. Williamsburg, VA: The Colonial Williamsburg Foundation, 1985.

Casselman, Bill. *Canadian Food Words: The Juicy Lore and Tasty Origins of Foods That Founded a Nation.* Toronto: McArthur and Company, 1998.

Collins, Louisa. [McClure, Dale, ed.] *Louisa's Diary: The Journal of a Farmer's Daughter, Dartmouth, 1815.* Halifax: Nova Scotia Museum: Nimbus Publishing, 1989.

Conrad, Margaret, et al, eds. *No Place Like Home: Diaries and Letters of Nova Scotia Women, 1771–1938.* Halifax: Formac Publishing Co., 1988.

Cormier-Boudreau, Marielle and Melvin Gallant. *A Taste of Acadie* [1978]. Fredericton, NB: Goose Lane Editions, 1991.

*Hearth and Home*

Crump, Nancy Carter. *Hearthside Cookery: An Introduction to Virginia Plantation Cuisine Including Bills of Fare, Tools and Techniques, and Original Recipes with Adaptations for Modern Fireplaces and Kitchens.* McLean, VA: EPM Publications, 1986.

Cuff, Robert, and Derek Wilton, eds. *Jukes' Excursions: Being a revised edition of Joseph Beete Jukes' "Excursions In and About Newfoundland During the Years 1839 and 1840"* St. John's: Harry Cuff Publications, 1993.

Daigle, Jean, ed. *Acadia of the Maritimes, Thematic Studies from the Beginning to the Present.* Moncton, NB: Chaire d'etudes Acadiennes, Université de Moncton, 1995.

Driver, Elizabeth. "Canadian Cookbooks, 1825–1949: In the Heart of the Home" in *Petits Propos Culinaires*, March 2003, 72: 19–39.

Duncan, Dorothy, et al, eds. *Consuming Passions: Eating and Drinking Traditions in Ontario.* Toronto: Ontario Historical Society, 1990.

Dunton, Hope. *From the Hearth: Recipes from the World of 18th-Century Louisbourg.* Sydney, NS: University College of Cape Breton Press, 1986.

Ennals, Peter and Deryck W. Holdsworth. *Homeplace: The Making of the Canadian Dwelling Over Three Centuries.* Toronto: University of Toronto Press, 1998.

Eustice, Sally. *History from the Hearth: A Colonial Michilimackinac Cookbook.* Mackinac Island, MI: Mackinac State Historic Parks, 1997.

Firth, Edith. *The Town of York, 1793-1815, A Collection of Documents of Early Toronto.* Toronto: Champlain Society and University of Toronto Press, 1962.

Glasse, Hannah. *The Art of Cookery, made Plain and Easy* (London, 1747). Facsimile edition. Totnes, Devon, UK: Prospect Books, 1995.

Greer, Allan. *People of New France.* Toronto: University of Toronto, 1997.

Guillet, Edwin. *Pioneer Inns and Taverns.* Toronto: Ontario Publishing Co., 1958.

Haight, Caniff. *Country Life in Canada* [1885]. Belleville, ON: Mika Publishing Co., 1986.

Healey, W.J. *Women of Red River: Being a Book Written from the Recollections of Women Surviving from the Red River Era* [1923]. Winnipeg: The Women's Canadian Club, 1967.

Jackel, Susan, ed. *A Flannel Shirt and Liberty: British Emigrant Gentlewomen in the Canadian

## Selected Bibliography

West, 1880–1914. Vancouver: University of British Columbia Press, 1982.

Lafrance, Marc, and Yvon Desloges. *A Taste of History: The Origins of Québec's Gastronomy*. Canada Parks Service and Les Éditions de la Chenlière, 1989.

Langton, Ann. [H.H. Langton, ed.] *A Gentlewoman in Upper Canada, The Journals of Anne Langton*. Toronto: Irwin Publishing, 1950.

Langton, John. *Early Days in Upper Canada: Letters of John Langton from the backwoods of Upper Canada and the Audit Office of the Province of Canada*. Toronto: MacMillan, 1926.

MacDonald, Eva M., "How the Cooking Stove Transformed the Kitchen in Pre-Confederation Ontario," in *Culinary Chronicles, The Newsletter of the Culinary Historians of Ontario*, Winter 2005, 43: 3–11.

Minhinnick, Jean. *At Home in Upper Canada* [1970]. Erin, ON: Boston Mills Press, 1994.

Moodie, Susanna. *Roughing It in the Bush; or, Life in Canada* [1852] and *Life in the Clearings versus the Bush* [1853]. Toronto: McClelland and Stewart, 1989.

Nightingale, Marie. *Out of Old Nova Scotia Kitchens*. Halifax: Petheric Press, 1970.

O'Brien, Mary. [Audrey Saunders Miller, ed.] *The Journals of Mary O'Brien, 1828–1838*. Toronto: MacMillan of Canada, 1968.

*Oxford Encyclopedia of Food and Drink in America*. [Andrew Smith, ed.] 2 vols. New York: Oxford University Press, 2004.

Roberts, Robert. *The House Servant's Directory* [1827]. Old Saybrook, CT: Applewood Books, 1993.

Silverman, Eliane Leslau. *The Last Best West: Women on the Alberta Frontier, 1880-1930* [1984]. Calgary, AB: Fifth House, Ltd., 1998.

Stewart, Frances. *Our Forest Home: Being Extracts from the Correspondence of the Late Frances Stewart*. Toronto: Gazette Printing, 1902 ed.

Traill, Catharine Parr. [Michael Peterman, ed.] *Backwoods of Canada: Being Letters from the Wife of an Emigrant Officer* [1836]. Ottawa: Centre for Editing Early Canadian Texts, 1997.

Traill, Catharine Parr. *The Female Emigrants Guide, or Hints on Canadian Housekeeping*. Toronto: Maclear, 1855.

Traill, Catharine Parr. [Rupert Schieder, ed.]

*Canadian Crusoes: A Tale of the Rice Lake Plains* [1852]. Ottawa: Carleton University Press, 1986.

Walker, Kathleen. *Ottawa's Repast: 150 Years of Food and Drink*. Ottawa: *The Ottawa Citizen*, 1995.

Whitelaw, Marjory, ed. *The Dalhousie Journals*. Vol. 1. Ottawa: Oberon Press, 1978.

# Photo Credits

L=left, R=right, B=bottom, T=top

The images on the following pages were photographed by Vincenzo Pietropaolo at Black Creek Pioneer Village, Toronto and Region Conservation Authority. The publisher wishes to thank the site and its curatorial and interpretive staff for their co-operation and support during the photography at Daniel Stong's first and second houses, the Half Way House Inn, and Burwick House.
3, 5, 6, 7, 9T, 9B, 12R, 13T, 16, 17R, 18B, 19T, 19B, 20, 21, 23T, 23B, 24B, 25T, 25B, 26, 27T, 27B, 28T, 28B, 29L, 29R, 30L, 31T, 34, 35T, 38, 39, 40, 41T, 43, 44, 49T

Other photographs were supplied by and appear courtesy of the following historic sites.

Colborne Lodge, Toronto, ON: 30R
Fort William Historical Park, Thunder Bay, ON: 37, 24T
Hutchison House Museum, Peterborough, ON: 11
Ste Marie among the Hurons, Midland, ON: 13B
Tourism New Brunswick:
Kings Landing Historical Settlement, Kings Landing, NB: 10,12L, 14, 17L, 18T, 22, 36, 41B, 42T, 42B, 49BR, 50
Old Government House, Fredericton, NB: 31B
Sherriff Andrews House, St. Andrews, NB: 49BL
Village Historique Acadien, Caraquet, NB: 15

The photographs on the following pages were reprinted with permission from the publisher:
*Fort Henry: An Illustrated History*, by Stephen D. Mecredy, photograph by J. Chiang, © 2000 James Lorimer & Co Ltd: 47
*Kings Landing* by George Peabody, photographs by H.A. Eiselt © 1997 Formac Publishing: 8, 24T, 45
*Louisbourg*, by Susan Biangi, photos by Norman Munroe, © 1997, 2001 Formac Publishing: 32, 33T, 33B, 35B, 46, 48

# Index

## A

Acadia, Acadians, 13, 14–16, 56, 57
Acadian Odyssey National Historic Site / Monument Lefebvre National Historic Site, 56
alcohol consumption, 30
Allison, Susan, 55
apple cellar, 24
Army
    officers' meals, 47–48
    rations, 51–52
    soldiers' meals, 57
ash, wood, 8

## B

backwoods cooking, 17–22
bake kettle (lidded iron pot), 11, 21, 22, 24
bake oven, 25–26, 55
baking, 9–10, 18
baking powders, 28
barbeque, 49
Barss, Beulah, 55
Battlefield House, 58
Battle of York, 48
Beavan, Frances, 54
bees, community work, 30, 53–54
berries, 16–17
Black Creek Pioneer Village, 25, 57
boiling, 9, 18, 29
bonfires, 49

*bouillon*, 15, 16
*brazier*, 33
breakfast, 20, 52
broiling, 9
butter, 18

## C

café, 46
caterers, professional, 45
cauldron (soup pot), 11
Champlain, Samuel de, 46
*chaudière* (soup pot), 10, 16
cheese, 18
chefs, 33–34, 37, 43, 45–46
Christmas Desserts by Lamplight, 58
Christmas dinner, 51
class
    gentlewomen, 21–22
    and standard of living, 31–35
coal, 8, 20, 50
coffee, 18, 20
Colborne Lodge, 57
Collins, Louisa, 17
confectioner *(confiseur)*, 45
cookbooks, 28
cooks, hired, 33, 36, 37–38, 43, 44–45
*coquemar* (soup pot), 10, 16
Craig, Sir James, 46
Cuff, Robert, 53

## D

Dalhousie, George Ramsay, ninth earl of, 52
Desloges, Yvon, 52
Diggle, Mrs. Carl, 55
dinner, 52, 54, 55
dinner (midday meal). *See* lunch
dishpan, 27
Doon Heritage Crossroads, 60
down-hearth cookery. *See* open-hearth cookery
Dundurn Castle, 35, 58
Dutch oven (lidded iron pot), 11

## F

Feltoe, Richard, 51
*les filles du roi* (the king's daughters), 13
firepit, 8
fireplace, 15, 19, 20
    gathering place, 24
    homesteads, 24
    laying of the fire, 19–20
    varied uses, 7
firewood, 20
First Peoples, 16, 17
fluted moulds, 7
food vendors, 41–42
Fort Chambly National Historic Site, 57
Fort George, 58
Fort Langley National Historic Site, 60
Fortress of Louisbourg, National Historic Site, 56

# Index

Fort William Historical Park, 59
Fort York, 57
*fourneau portager* (portager stove), 33
Franco, Rossi, 46
French-Canadian cuisine, 15–17, 54
*fricot*, 15
frying, 9, 18

## G

gendered division of labour, 8, 13, 17–18
German cooking, 17
German emigrants, 13, 18, 20, 24
Gibson House, 57–58
girdle, 11
Glasse, Hannah, 28
goose wing, 39
griddle, 11

## H

Hargrave, Letitia, 21
Hogmanay, 58
homes
    Acadian and Canadien rural homes, 14–15
    expanded homes, 23–25
    large town/country house, 35–36
    manor houses, 32–33
    one-room cabin, 14, 19
hominy, 8
Horton, John, 46
hot hearths. *See* portagers
house-raising bee, 53–54
Howard, Jemima, 57
Howard, John, 57
Hutchison House Museum, 59

## I

Indian corn, 16, 22
inns and taverns, 42–45
Ireland House at Oakridge Farm, 59
Italian immigrants, 46

## J

Jarvis, Hannah, 48
Jukes, Joseph Beete, 53

## K

Kalm, Pehr, 43
kettle (soup pot), 10
King, Mackenzie, 61
Kings Landing Historical Settlement, 56
kitchen help, 28, 36
kitchens, 15, 19, 23–24, 33, 35, 38–40
kitchen utensils, 26–27, 36

## L

Lafrance, Marc, 52
Langlois, Charles-René, 46
Lang Pioneer Village, 59
Langton, John, 20
La Varenne, François Pierre, 45
Liger, Louis, 45
Louck's Farm House, 60
Lower Fort Garry National Historic Site, 59
Loyalists, 13–14, 52, 56
lunch, 53–54

## M

Mackenzie House Museum, 60
mail-order catalogues, 40
Mair, Charles, 54
maple syrup, 16, 17
mason jars, 29
McCowan Log House, 58
McNab, Sir Allan, 35
Mennonites. *See* German emigrants
Montgomery's Inn Museum, 44, 58
Moodie, Susanna, 18, 21

## N

New Year's dinner, 51
Norquay, Elizabeth, 30

## O

Oakridge Farm, 59
O'Brien, Mary, 22, 44
open-hearth cookery, 8–11, 15, 56, 57
    burns, 39
    cooking techniques, 9–10
    cookware, 10–11, 16
    implements, 8–9
    intricacies of, 20–21, 22, 29–30, 37–38
*l'ordre du bon temps* (The Order of Good Cheer), 46
Ormsby, Margaret A., 55
oven, 25–26, 27, 49, 55, 57

## P

*paillase*, 33
pastry cook (pâtissier), 45
pearlash, 8
pemmican, 47
Pickering Museum Village, 59
pickles, 17, 28–29

pie safes, 39
po tagers, 33, 36
potholders, 9
pots, 10–11, 18, 21
*poutines,* 16
preserves, 17, 28–29
puddings, 18, 29–30

## Q
Quebec Driving Club, 52

## R
Ramsay, George. *See* Dalhousie, George Ramsay, ninth earl of
ready-made food, 40
reflecting oven, 27
roasting, 9, 16, 27
Roberts, Robert, 38
Rollet, Marie, 13
Ross Farm, 57
Rundell, Eliza Maria, 28

## S
Ste. Marie among the Hurons, 59
Scarborough Historical Museum, 58
Schneider Haus, 60
scullery girl, 36
Simcoe, John Graves, 34
Simcoe, Lady, 34
smokehouse, 24
smoking, 8
soft cheeses *(schmierkäse),* 18
Spadina Museum: Historic House and Gardens, 60–61
spit-roasting, 9

Stewart, Ellen, 20
Stewart, Frances, 7, 18–19, 22, 53
Stewart, Thomas, 53
stew holes. *See* portagers
stove, iron, 40, 50
Strong, Daniel, 24
Strong, Elizabeth, 24
sugar, 29
supper, 54
surveying trail, meals, 53

## T
taverns. *See* inns and taverns
tea (beverage), 20, 54
tea (meal), 54
*terrine* (baking dish), 16
Thompson Memorial park, 58
tinware, 26–27
Todmorden Mills: Heritage Museum & Arts Centre, 58
*tourtes,* 16
*tourtière* (covered pan), 16
Traill, Catharine Parr, 22
trivets, 8

## U
Ukrainian Cultural Heritage Village, 59–60
Upper Canada Village, 59

## V
Verchères de Boucherville, Thomas, 41
Le Village de l'Acadie, 57
Village Historique Acadien, 56

## W
water pump, 27

Wentworth, Sir John, 34
Wentworth, Lady, 34
Westfield Heritage Village, 59
Whitelaw, Marjory, 52
wild rice, 19
Willcock, Joseph, 51
Wilton, Derek, 53
women
　and cooking, 8, 21–22, 30, 41–42, 43, 48
　domestic skills, 21–22, 28, 32
　housewifery, 11–12, 17–18, 19–20
　immigration, 13–14
　paid work, 28, 32, 33, 36
wood-fired oven, 49
Woodside National Historic Site, 61